Geometric Principles and Procedures for Computer Graphic Applications

Geometric Principles and Procedures for Computer Graphic Applications

SYLVAN H. CHASEN

Lockheed-Georgia Company

Prentice-Hall, Inc., Englewood Cliffs, New Jersey 07632

Library of Congress Cataloging in Publication Data

CHASEN, S. H. (date)
  Geometric principles and procedures for computer
graphic applications.

  Bibliography: p.
  Includes index.
  1. Computer graphics. 2. Geometry, Analytic.
  I. Title.
T385.C46   001.6′443   78-7998
ISBN 0-13-352559-7

©1978 by Prentice-Hall, Inc., Englewood Cliffs, N.J.   07632

Printed in the United States of America

10   9   8   7   6   5   4   3

PRENTICE-HALL INTERNATIONAL, INC., *London*
PRENTICE-HALL OF AUSTRALIA PTY. LIMITED, *Sydney*
PRENTICE-HALL OF CANADA, LTD., *Toronto*
PRENTICE-HALL OF INDIA PRIVATE LIMITED, *New Delhi*
PRENTICE-HALL OF JAPAN, INC., *Tokyo*
PRENTICE-HALL OF SOUTHEAST ASIA PTE. LTD., *Singapore*
WHITEHALL BOOKS LIMITED, *Wellington, New Zealand*

To my family—
Cathie, Debbie, Diane, Jane, and Susan—
for their unflagging support.

CONTENTS

Chapter III
THREE-DIMENSIONAL GEOMETRY *125*

APPENDICES *193*

BIBLIOGRAPHY *223*

INDEX *235*

FOREWORD

Sylvan "Chase" Chasen, is one of the authentic pioneers in computer graphics. I'm delighted and honored that he asked me to write this foreword to his latest contribution in the field.

So often we get caught up in the hardware and software details of a system, particularly computer graphic systems where such details often determine if the final result is usable or not, that we lose sight of the fundamental assumptions around which many of the systems are formulated. One of these fundamental assumptions in computer graphics is geometry, ranging from conventionally understood aspects of geometry as related to lines, curves, and drawing construction, to the less often understood implication of geometry, related to curve fitting and data presentation and principles. Chasen has done a superb job of bringing together the variety of geometric ideas which underlie much of computer graphics. The material he includes is not readily found in any other single source. The treatment is lucid and readily understandable by the intended audience. Mr. Chasen, early on, states that there are really two audiences for the book. One is the systems programmer, who must give the systems user a variety of geometric techniques to help the user solve the problem. The other is the user, who basically blindly accepts the techniques as tools to help in solving the problem. The analyst needs the data to program the tools—the user needs to understand the basis of the program in order to get a sense of the accuracy and limitations of the tools. I know of no one better qualified to talk about the basics and the implications of these basics than Mr. Chasen.

In addition to the basic geometric discussions, Mr. Chasen brings his broad experience to bear in evaluating the practical aspects of the various choices—the questions of accuracy, the question of which method is better suited to different display technologies, questions of how the user can graphically indicate his geometric requirements to the system. These are all concepts far beyond the simple geometric derivations, but are a part of the real world of graphics.

In all, Mr. Chasen does a most effective job in constantly relating the geometric principles to the question "What does all this have to do with Computer Graphics?" This is not a theoretical text about geometry. This is a contribution to the field of computers and graphics and the text constantly stresses how the presented data relates to the environment in which it is to be used.

Geometric Principles and Procedures for Computer Graphic Applications is a significant contribution to the fundamental body of literature which lays the foundation for continuing growth in the field.

CARL MACHOVER
President
Machover Associates Corp.

ACKNOWLEDGMENTS

In the preparation of this text, I have received support and constructive suggestions from more people than I can adequately acknowledge. It is most sincerely appreciated.

I want to express particular appreciation to Carl Machover and Alan Adams whose encouragement and assessment of the need for the text were instrumental in my undertaking the task of its development.

Mrs. Jo Atcheson has had the responsibility of all the typing of the manuscript and the related functions of composition. Her superlative performance has significantly shortened the time from the concept of the text to its publication. John Osterman and Dave Prince were of considerable help in preparing the illustrations and the index, respectively. Also, I would like to express my appreciation to Bill Gordon, Robin Forrest, Nelson Logan, and Rich Riesenfeld for their furnishing me with listings useful in the preparation of the bibliography.

Finally, I want to express my thanks to the Lockheed Corporation. Though I had a severe visual impairment, Lockheed-Georgia offered me a position in 1958. Their support of qualified persons irrespective of physical or social status was in evidence well before open-mindedness in employment became a more comprehensive national objective. This enlightened environment has been of immeasurable value to me in my attempt to develop and share technical experiences.

Figures 14, 15, 42, 43, 65, and 74 are courtesy of Lockheed-Georgia Company.

Sylvan H. Chasen

underway to integrate functions within major fields, such as engineering and manufacturing; although comprehensive integration and interfacing between major fields is still in a primitive state.

In the attempt to reach the lofty objectives of systems integration, we recognize some of the subordinate objectives to be the development of compact data, computational ease, better interfaces to facilitate further analysis, graphic visualization for better man-machine interaction, timely hard-copy output, efficient data storage and retrieval, and so forth. The number of hardware/software configuration options are virtually unlimited; so the achievement of the highest levels of efficient integration is a distant target. Its approach is more dependent on the artistry of human developers than on scientific optimization. There is no doubt that the careful mathematical and geometric formulation of data, of relationships, and of design criteria will enhance the major integration objective through the enhancement of each of the subordinate objectives.

The purpose of the following text is to present what I consider to be many of the most important basic geometric considerations that a proponent or prospective user of interactive graphics or a designer/analyst should be aware of. It is not intended to be as detailed as some texts, but enough detail will be given to serve as a comprehensive instructional base and a useful reference and/or courses in applied graphics and mathematics. It will cover and expand on many of the concepts set forth by Rogers and Adams.[1] The objective here is to organize the discussion so that it reflects more of a practical vent—leaving much of the more basic detailed derivations for the reader to complete as required or locate in other textual material, particularly that just mentioned. However, formulas and procedures that apply to each of the considered techniques will be given, and certain derivations not normally found in the literature will be presented. Examples to illustrate the use of all formulas and procedures will be given. Along with the description of the techniques and the corresponding formulas will be a discussion of the philosophy and principles of applicability. There is no intention to set forth application procedures or algorithms, such as those in the literature for numerical milling, printed circuit boards,

[1]David F. Rogers and J. Alan Adams, *Mathematical Elements for Computer Graphics* (New York: McGraw Hill, 1976).

and stress analysis, but to set forth the primary geometric bases from which applications are built. A large number of the most common and most versatile of such bases will be treated. Exercises are given at the end of most sections to enhance study and understanding. Geometry, as used here, pertains to curve and surface generation. Some important transformation formulas are presented but general transformation theory, geometric projections, and hidden line and surface techniques are not treated within the scope of this text, as they are aptly covered in other texts.When studying the various math-model options, it is important for the reader to realize that the console user need not necessarily understand the math in order to apply the techniques. The equations can be set up *automatically* and solved *automatically*. This is the province of the systems programmers and analysts who must provide the graphic design and analysis tools. The user should understand the suitability of the various options that are provided. User actions require the input of constraints, such as points, slopes, and other data. Of course, a comprehensive understanding of the principles will greatly benefit a user by helping to define application processes and procedures.

The discussion will be broken into three logical divisions as follows:

1. Displaying existing equations or mathematical forms.
2. Creating a mathematical formulation to satisfy known or desired data constraints.
3. Three-dimensional geometry.

1

DISPLAYING
EXISTING EQUATIONS

For many applications amenable to computer graphics, there is little need to develop any kind of math model. It is merely necessary to display a series of data points (individually or connected by straight lines), to display characters, or to display equations that must be converted into a series of vectors (straight lines) over a specified interval or range.

When vectors and/or characters are explicitly required, the need may be fulfilled by built-in hardware function generators. Vector and character generators are two of the most commonly used function generators in computer graphic hardware. Thus, such generators essentially automate the line and character generation functions. When an equation is to be represented, there are certain options that might be an integral part of the system software. In particular, the nature of the application may influence the way vectors should be generated to meet application objectives—objectives that may be more esoteric or more functional. For example, many basic applications require nothing more complex than the definition of a circle or a circular arc; so we will use this shape to discuss the principles of describing a sequence of vectors. Generally, only elementary action is required by the terminal operator to create circular arcs. The operator may specify the center and radius that will suffice for the setup of a circle description. There may be a hardware circle generator to automatically produce the circle, or it may have to be done by software—preprogrammed by a software supplier or by the using installation. In either case the nature of the application, the amount of total display drawing, and the type of display hardware should influence what process will be provided by the software circle routine. The basic circle-defining equation of

$$(X - h)^2 + (Y - k)^2 = r^2 \qquad (1)$$

indicates that (after (h, k) for the center and r for the radius are input) the circle can be plotted by incrementing X in intervals between h and $h + r$ (symmetric properties alleviate the need for using the interval h to $h - r$, also). The number of intervals may be set in the routine. Each selection of X in the interval leads to two values of Y and therefore two pairs of coordinates on the circle that may be plotted on the display. Taking successive pairs of coordinates within the data set and instructing (within the routine) the vector generator to draw lines from point to point will give a vector approximation of the circle that is to be represented. Several hypothetical questions and considerations arise. Based on the incremental selection of X's, are there enough points to portray the circle adequately? Does the nonuniform spacing of points present a problem (since Y values are not equi-spaced)? Is there a sufficient amount of total computation requirements (for many circles on the display) to be concerned with the method of computing coordinates or to take into consideration the type of CRT on which the application is displayed? With regard to the first question, if the same number of points is used for all circles (say 24), there may be more points than needed for circles of "small" radii and fewer than needed for circles of "large" radii in order to get an acceptable portrayal. Furthermore, excessive points displayed too close together may have the disadvantage of portraying a nonpleasing picture, since some CRT's have poor closure at designated points; that is, end points do not mate well, and this will be more evident if points are very close and are of high relative density.

In some applications either aesthetics or practical considerations will dictate the requirement for equi-spaced points. For example, it may be necessary to display a circular arc tangent to a line. Since the circle is broken up for display into many short lines (vectors), the console operator may prefer the depicted arc to, at least, give the general appearance of tangency on the display screen.

Equi-spacing of points could benefit the visualization and hence the feeling of confidence that the picture on the display is a model of what the console operator wants to describe within the computer data base. Borrowing from the next chapter on creating math formulations, the need for equi-spacing can be accomplished by using parametric techniques. That is, points on the curve (in this case the circle) can be derived by the computer routine such that the chord lengths (which approximate arc lengths) are the same from point to point. The X and Y coordinates of such points are then portrayed on the display.

For many applications the computation time needed to develop the set of points to be displayed is of little concern. However, if much data is to be displayed and if the CRT is of *refresh* type, it is possible to run into a *flicker* problem. This occurs when the beam deflection circuitry cannot present the entire picture fast enough to enable the eye to see a constant picture. Time for computation and display can often be saved with no penalty in programming complexity. One way is to use, when possible, fewer points to represent the curve. In the case of a circle, there may be a fixed number of points to be generated for each circle, or one might program the system such that the number of points is related in some prescribed way to the radius. For curves in general, step sizes may be derived as a function of curvature (which can be derived from the known equation) and the acceptable tolerance between the vector representation of the curve and the true curve. Such decisions depend on programming preferences, overhead to accomplish the added computations, and the application requirements. A way to acquire equal spacing and also cut down on computation time can often be found. For a circle, a recursive relationship for X and Y as a function of θ may be developed from the parametric equations

$$X_N - h = r \cos \theta$$
$$Y_N - k = r \sin \theta$$
$$X_{N+1} - h = r \cos (\theta + d\theta)$$
$$Y_{N+1} - k = r \sin (\theta + d\theta)$$

(2)

where h and k are the coordinates of the center, (X_N, Y_N) are the coordinates of any point, θ is the angle between the X axis and a line from the center to that point, $d\theta$ is the constant angular increment, and (X_{N+1}, Y_{N+1}) are the coordinates of the next point.

Formula (2) may be expanded and simplified to give

$$X_{N+1} = h + (X_N - h) \cos d\theta - (Y_N - k) \sin d\theta$$
$$Y_{N+1} = k + (X_N - h) \sin d\theta + (Y_N - k) \cos d\theta \tag{3}$$

Thus, the circle can start from any arbitrary point, and each successive point can be computed recursively with equal spacing. This can be done by computing $\sin d\theta$ and $\cos d\theta$ once and thus eliminate computation of trig functions for each point. For example, suppose the initial point is taken arbitrarily (no loss in generality) at $\theta = 0°$. Then the first point (X_1, Y_1) is $X_1 = h + r$ and $Y_1 = k$. Then (X_N, Y_N) for all points is obtained by Eq. (3), in which each point's coordinates are derived from known data and the coordinates of the previous point. Thus, minimal computation is required and equi-spaced points are achieved. An alternative parametric technique for generating the circle is

$$X = h + r\left(\frac{1 - t^2}{1 + t^2}\right), \qquad Y = k + r\left(\frac{2t}{1 + t^2}\right) \tag{4}$$

for $0 \leqslant t \leqslant 1$ (also using symmetrical points). This is somewhat of a compromise between speed of computation (no square roots or trig functions) and equal spacing of points (not equal spacing, though more so than simply incrementing X and solving for Y).

Where many circular or elliptic arcs characterize an application, transformation can be applied to a *standard* circle to save considerable computation. We might, for example, compute a set of coordinates to represent a circle with a unit radius and its center at the origin, $X^2 + Y^2 = 1$. If we wish to portray another circle with its center at (h, k) and radius r, we need only to scale both X and Y for each data point by the factor r and then translate each coordinate pair by h and k, respectively. This process can be repeated for all circles as long as we do not require a change in the number of describing points. If we should wish to display an ellipse or portion of an ellipse, we need only ascertain the lengths of the semimajor and semiminor axes and the angle between the major axis and the X axis. This can readily be derived from standard algebraic treatment of the describing elliptic equation. Then it is merely necessary to scale each X of the unit circle by the length of the semimajor axis, each Y by the length of the semiminor axis, and rotate the resulting data points by the appropriate angle. Transformations for scaling, translation, and rotation may be conveniently prescribed via conventional matrix operation or by matrices using homogeneous coordinates. The principle of using a basic shape with certain parameters of scale, rotation, and so forth, can be applied to many common constructions. The basic shape is often called a *primitive*. Homogeneous coordinates and matrix operations are described in many places.[1]

Should the CRT be a *storage* tube rather than a *refresh* tube, the speed of computation and display of all desired data will not be a factor regarding flicker. The tube will hold everything that is traced out. The principle factors that should be considered for storage displays are the transfer rates of data to the display and the frequency of changes in data that are to be displayed. Of course, this relates to the nature and extent of the application. Thus, the CRT system and the applications that use it should influence the choice of programming options to ensure operational efficiency and minimum cumulative costs for software development, computer operation, and manpower. In this regard preliminary and updated planning will generally be very rewarding.

Another important consideration in displaying data is the translation of the displayed picture to a hard-copy picture or plot. In many cases it is sufficient to take a photograph of the display—conventional or Polaroid-type. When a larger or more accurate portrayal than that from a photograph is desired, there are two alternatives. First, the coordinates that represent end points of vectors on the display can be used to produce a plot on any number of plotters. The most expedient output of this type is without change in scale; that is, what you see on the display is exactly what you get on the plotter—vector for vector and dimension for dimension. In some cases of this kind, a scale change may be requested to derive a larger hard-copy plot of what is displayed. That is quite straightforward as long as there is no need to change the actual vector end coordinates (only scale change). If such vectors are increased in length to satisfy scaling, the plot of the picture could be aesthetically displeasing. Also, no increase in precision will be

---

[1]See, for instance: David F. Rogers and J. Alan Adams, *Mathematical Elements for Computer Graphics* (New York: McGraw Hill, 1976). Newman and Sproull, *Principles of Interactive Computer Graphics* (New York: McGraw Hill, 1976). S. H. Chasen, "Applied Interactive Computer Graphics" (lecture notes, UCLA short course, 1973–77).

achieved. This leads to the consideration of the second alternative, which is needed to produce a hard-copy plot that is larger than the display and is aesthetically pleasing and/or has increased precision. This requires:

1. The fitting of data with some mathematical model or the nonlinear interpolation of the data to get a more precise portrayal of intermediate data. (The reader should note that the breaking up of a given vector into a series of shorter vectors will not serve any positive purpose because this would yield more points, all of which will be along existing straight lines with no improvement in curve representation.)
2. The existence of an appropriate describing equation or algorithm within the computer that can be used to extract enough data points to achieve a prescribed level of precision. Procedures have been developed and programmed by suppliers and users such that certain equations can be represented by a series of vectors to meet specific tolerances when plotted at any desired scale. These procedures depend primarily on the length of a chord of the exact curve and the curvature of the curve as derived from the describing equation. Typically, this kind of data interpolation exists for polynomials, splines, and conics; although it can be formulated for any type of mathematical model.

The discussion up to this point merely outlines some of the more important factors of consideration in addressing the first question—that of displaying existing data and/or equations. Though a circular arc was used for illustrative purposes, the principles set forth apply, in general, to all types of curves. The display technique from a programmer/analyst point of view is, after all, a systems problem when attacked with thoroughness.

The actual technique for deriving the equation from given constraints will be discussed in the next chapter. Transformations to display coordinates, clipping and windowing techniques, and other important special treatments of data that have been well covered in the literature are not considered as part of the scope of this text.

II

CREATING A MATHEMATICAL FORMULATION TO MATCH KNOWN OR DESIRED DATA CONSTRAINTS

The fields of data reduction, analysis, and design have one requirement in common—representing data by mathematical formulas. In the process of developing such formulations, it is desirable (if not necessary) to consider the factors of computational ease, desired compactness of data, further analytical treatment of the representations, and visualization as characterized by graphic displays, hard copy drawings, microfilm processing, and so on. In many cases, the underlying mathematical form is known from the nature of the problem: as examples, seasonal temperature variation may be sinusoidal; a trajectory in space or a uniformly loaded bridge cable is a polynomial of second degree; a powered missile may be represented by a third order or higher degree polynomial because of changes of accelerations; decibels of sound are logarithmic; growth phenomena or compound interest is exponential; the relationship between pressure and volume at a given temperature of a gas is hyperbolic; satellite orbits are any of the mathematical conics; and so forth. When the nature of the physical phenomenon is known, it may be natural to fit empirically derived data with the appropriate underlying and descriptive mathematical function or functions. In many cases the underlying function is either unknown or in doubt. Then it is natural to ask the question of the analyst, "What kind of fit or representation do you want?" This type of question is often met with perplexity, if not frustration. A common reply is, "I want a smooth curve." That statement logically leads to the follow-up question, "What do you mean by smooth?" The retort to that question usually involves considerable arm waving and a lot of verbiage that still falls short of answering the question. The truth is that *smoothness* means different things to different people and generally relates to a personal judgment of intangible factors. It may be purely aesthetic, that is, the curve should *look* smooth, which, of course, has little or no analytical meaning as far as formulation is concerned. Of course, smoothness might be defined in terms of the continuity of derivatives of various orders, the passing through all points with some kind of spline, or some type of *least squares* statistical fit; or it might be defined by the averaging of the ordinates of several points over a short span. The averaging technique is a good candidate for smoothing when the independent variable has uniform spacing (e.g., equal increments of time), and the dependent variable has an approximately linear trend over the span. The average of a span represents the smoothed center point of the span and is plotted there; then the span is shifted by one point. The process of averaging and plotting is repeated. Thus, we might average points one through five and plot the average as the new or smoothed point three; average points two through six and plot as the smoothed point four; and so on. This process does not accommodate the end points of a data set, but this is usually of little or no practical concern. If it is, a variety of simple procedures can be developed. If the criteria for equal spacing and linearity can be reasonably assumed, more smoothing is achieved for larger spans. This comes about from statistical theory. If the standard error of measurement of a single point is S_E, the average of an N point span has a standard error of $S_E / \sqrt{N}$. Thus, for five point spans, the data points are smoothed by reducing the scatter from measurement error by a factor of $1/\sqrt{5}$. Individual errors are, therefore, reduced to less than $1/2$ the original error when each ordinate is replaced by five point averages. Variable spans and other variations of this technique can easily be developed for use with interactive graphics.

An example of an elementary type of smoothing that has aesthetic virtues is illustrated by one of the earliest experiments in graphics. A console operator may *draw* a free-hand curve or write his name on the display. This is done by the simple expedient of having the computer set a graphic symbol or cursor (tracking cross, enlarged dot, spiral, or what have you) at some initial location. As the operator moves it with a *tracking* device, the computer will continually update the distance from its present location to the previous location. When this distance equals or exceeds a small programmed or input constant, a short vector will be created between the two points. As the cursor continues to be moved, successive connecting vectors will be generated. The small constant (just a few grid units) ensures a remarkably good representation of the curve or name that the operator would have drawn or written on paper. It includes all the wiggles that the person would normally make through involuntary manual movements. Because the vectors are tiny, the depiction appears to be a continuous curve or approximately so. The prescribed smoothing or filtering process was as simple as one could imagine—on operator command (push an appropriate button) to skip every other point. That is, connect the first and third, third and fifth, and so forth, and eliminate the intermediate points. When this is done, the hand drawing will actually look smoother even though the vectors are longer. When the process is repeated (connecting every other point), continued improvement in aesthetic appearance will occur until some ultimate is reached. Then, as the process continues, the picture begins to look boxy and less pleasing. That is because the lines become noticeably long. The number of applications of the smoothing process necessary to achieve maximum apparent smoothness depends on the degree of ineptness, wiggles, and so forth, in the starting hand-drawn curve. The perceived smoothness of the set of resulting curves will be in the eye of the beholder. It will be in the gray area of human experience, preference, and pattern recognition —faculties which can be well served by interactive graphic dialogue.

We will now proceed to develop the basic characteristics of many of the commonly used curve and surface-fitting techniques.

A. LEAST SQUARES

In the least squares case, the problem of defining smoothness may be exacerbated by having to make the decision as to what type of math model will be used for the fit. Most typical models are polynomials. There are various principles that may aid the analyst in defining his concept of smoothness, but there is no general agreement as to what is wanted when a smooth fit is requested. To fit a large number of math functions of different types and to pick the *best* fit according to some criterion may have some benefit, but this tends to be extravagant with computational time. What is more important, and perhaps limiting, is that such an approach is often the result of trying to relegate the decision process to the computer by the maneuver of using a large number of math models. The problem with this approach (usually not well comprehended by the uninitiated) is that the number of such models, although seemingly large, is actually relatively small—certainly finite. There are ways in which graphics may be particularly valuable in obviating this problem.

1. Manually Derived Least Squares Fit

By inspecting data presented on a display, the display console operator can be given (via programming) the capability to move a tracking symbol and create a hand-driven (via the cursor) sketch of the curve that he would consider, according to his mental judgment, to be the representative fit of the data. This is along the lines of the example discussed just prior to this section. The curve would have small wiggles as a consequence of normal manual jitters and would not, at this juncture, be represented in the computer by a math model. An example of a hand-drawn smooth curve through a set of data points is shown in Fig. 1. Also shown is a comparison linear fit, which is discussed later.

Suppose that the 12 points of Fig. 1 have the following coor-

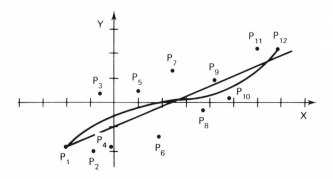

Figure 1 Set of points smoothed by a hand-drawn curve and by a linear least squares fit.

Point	X	Y	Y' (From Curve)
P_1	−2.1	−1.8	−1.8
P_2	−0.9	−2.0	−0.9
P_3	−0.6	+0.4	−0.7
P_4	−0.1	−1.9	−0.6
P_5	+1.0	+0.5	−0.2
P_6	+1.8	−1.4	−0.05
P_7	+2.4	+1.3	+0.1
P_8	+3.7	−0.3	+0.3
P_9	+4.2	+0.9	+0.4
P_{10}	+4.8	+0.2	+0.6
P_{11}	+6.0	+2.2	+1.4
P_{12}	+6.8	+2.1	+2.1

dinates, and that the Y's are the corresponding ordinates as measured on the manually-drawn curve.

The differences between Y' and Y represent the degree of smoothness in a subjective sense. The Y' might be regarded as the best fit for a set of empirical data. On a graphic display, one might wish to consider several alternative hand-drawn curves as candidates for the best fit, according to the nature of the application and operator judgment. Regardless of the curve, *goodness of fit* is often measured by a correlation coefficient, R. In essence, R is a measure of the usefulness of X as a predictor of Y as compared to that of the average of the set of Y's (irrespective of X) as a predictor of Y. In statistical terms, R^2 is computed by subtracting, from unity, the ratio of the variance of the Y data with respect to the fitted or assumed curve over the variance of the Y data with respect to the average of the Y's. Mathematically, this is written as follows:

$$R^2 = 1 - \frac{\frac{1}{n}\Sigma(Y_i - Y')^2}{\frac{1}{n}\Sigma(Y_i - \overline{Y})^2} \tag{5}$$

where Σ refers to

$$\sum_{i=1}^{n}$$

for Eq. (5) and for subsequent text.

If the curve were to pass through each point, the numerator of the fraction in Eq. (5) would become zero and R^2 would become unity. When the curve Y' does not improve the fit (in comparison to no fit), the numerator of the fraction would be the same as the denominator and R^2 would be zero. In other words, the closer R^2 approaches unity, the better Y' is in predicting Y for any X; and the closer R^2 approaches zero, the more useless Y' is in predicting Y. Using the data of Fig. 1 and Eq. (5), R^2 is computed to be approximately 0.63 or R is approximately 0.79. This means that Y' is better than $\overline{Y}$ (the average) as a predictor of Y. The reader should note that we have computed one measure of goodness of fit without having fit a specific mathematical function to the data set. Not only does graphics afford the opportunity to apply the more flexible hand-drawn fitting technique, but it also permits the operator the opportunity to peruse the data to eliminate points that he considers to be erroneous or wild. He may also add data according to his knowledge of the nature of the application and the judgment that goes with it. The pattern of points may reveal one or more points that abnormally depart from the rest. Whether hand fitting or analytic fitting (which we will discuss next) is in order, the meaningful use of correlation techniques is predicated on the use of more or less homogeneous data. Points that deviate from the mass of points carry more weight in least squares fitting than is warranted and should, therefore, be discarded from the computation of R^2, even if the points are deemed valid from other considerations.

2. Mathematical Least Squares (2-D)

If subsequent analyses indicate a need to derive a math model of a smooth curve, there are two alternatives. One is to select a set of points from the hand-drawn Y' and apply one of the curve-fitting options—polynomials of appropriate degree, splines, etc.—that will be explained later. The other alternative is the most common—the least squares fit of a mathematical equation of some kind to the data. For any such mathematical model, it would be necessary to solve for certain parameters of the equation type in order to minimize the sum of the squared deviations between the observed Y's and the fitted Y'. In general, this can be done by numerical trial and error or other methods. If, for example, we wish to use the form $Y' = a \sin bX$ to fit data in the least squares sense, we cannot derive a and b directly; but we can select trial values, compute the sum of the squared deviations, and continue the process toward the objective of finding the values of a and b that lead to a minimum sum of squared deviations. For certain mathematical forms we can actually solve for the parameters directly. This is true for any polynomial. Polynomial least squares fitting is commonly used because, not only is it easy to derive the coefficients, it also has other favorable qualities, such as known number of maxima and minima on the curve, potential for inflection on the curve if desired, ease of storage and retrieval, and ease of computation.

Suppose we wish to determine the best linear fit to a set of data. Then we must determine A_0 and A_1 such that

$$Y' = A_0 + A_1 X \tag{6}$$

and that the sum of the squared deviations between Y' and the Y's of the data points is a minimum. This has been derived in many standard statistical texts such that two linear equations in A_0 and A_1 will fulfill these requirements. Thus,

$$A_0\Sigma 1 + A_1\Sigma X_i = \Sigma Y_i$$
$$A_0\Sigma X_i + A_1\Sigma X_i^2 = \Sigma X_i Y_i \tag{7}$$

The summations range over n data points and are easy to compute.

Now suppose, as a numerical example, we have 12 data points as shown in Fig. 1, and we wish to get the least squares line. The resulting equations, which the reader can verify by performing the indicated summations, are

$$12A_0 + 27A_1 = 0.2$$
$$27A_0 + 152.2A_1 = 37.74$$

Solving for A_0 and A_1 gives

$$Y' = -0.92 + 0.41X$$

Then we compute R^2 using the appropriate formula where $\overline{Y}$ is $(-1.8 - 2.0 + 0.4 + \ldots + 6.8)/12$ and Y' is determined for the X's, that is, for -2.1, -0.9, etc. R^2 is found to be 0.60, which indicates that Y' is somewhat better in predicting Y as a function of X than is $\overline{Y}$ (which does not vary with X).

We note that the mathematical linear fit is almost as good, with respect to R^2, as the hand-drawn case. It is quite possible that the hand-drawn fit could be either considerably worse or considerably better than any of the polynomial-fitting options—depending on operator aptitude and the extent that the operator can justify drawing a wavy curve that comes close to the data points.

Had we used a second degree polynomial (quadratic), the least squares equation could be represented as

$$Y' = A_0 + A_1X + A_2X^2 \tag{8}$$

Then the A's could be derived from the set of linear equations given by

$$A_0\Sigma 1 + A_1\Sigma X_i + A_2\Sigma X_i^2 = \Sigma Y_i$$
$$A_0\Sigma X_i + A_1\Sigma X_i^2 + A_2\Sigma X_i^3 = \Sigma X_i Y_i \tag{9}$$
$$A_0\Sigma X_i^2 + A_1\Sigma X_i^3 + A_2\Sigma X_i^4 = \Sigma X_i^2 Y_i$$

The computation of the summations is, of course, straightforward. By inspection of the equations for the A's for the linear fit and the quadratic fit, and by simple extension, it is easy to write the formation equations for a polynomial of any desired degree. We are limited to a polynomial of $N - 1$ degree to pass through N points, for example, a cubic (third degree) polynomial least squares solution will pass through a set of four points; that is, the sum of the squared deviations with respect to Y' and R^2 would be zero and unity, respectively. This would not mean that the curve is a "perfect" fit (except at the data points), because there are no *degrees of freedom*. One should have some degrees of freedom to give meaning or significance to the correlation coefficient. For a polynomial fit of degree D to n data points, there will be $n - D - 1$ degrees of freedom. Discussion of degrees of freedom and its implications is found in most statistical texts. One simple way of understanding it is as follows: if we wished to fit a least squares line to two points, the line would pass through the points; but if there were three points, the line would not generally pass through all three. There would be one degree of freedom.

The higher the selected degree of polynomial fit, the closer the resulting curve will approach each point. However, the curve will wiggle to fulfill this effect. Therefore, for numerous data points there would be considerable wiggling, which in most cases is not desired. This is why the console operator must exercise judgment in selecting the appropriate degree if polynomials are used. Another way of citing the closer approach to the data points is that higher degree fits must have higher values of R^2. That is, a second order polynomial cannot produce a worse fit than a linear fit, since a linear fit is in the set of second-order polynomials—with zero coefficient of the quadratic term. The same principle holds for still higher order polynomials. Graphics can be of considerable assistance in determining the most practical order of polynomial fitting, since the magnitude of R^2 is *not* a criterion in itself. We can always make R^2 unity if the degree of fit is sufficient to make the curve pass through every point. To illustrate how least squares polynomial fitting approaches the data points with higher degree functions, we will use four very simple points, Eq. (9), and a four-point fit to a cubic (explained in the next section). Suppose the four points are $(-1, 0)$, $(0, 1)$, $(1, 0)$, and $(2, -2)$. The average of the Y's is, therefore, $\bar{Y} = -1/4$ (the best estimate of Y without relating Y to X). The least squares linear, quadratic, and cubic fits are, respectively,

$$Y' = \frac{1}{10} - \frac{7}{10} X$$

$$Y' = \frac{17}{20} + \frac{1}{20} X - \frac{3}{4} X^2$$

and

$$Y' = 1 - \frac{X}{6} - X^2 + \frac{X^3}{6}$$

Figure 2 shows the four points and the three polynomial least squares fits.

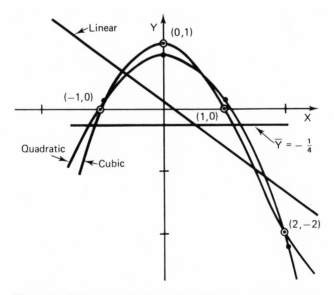

Figure 2 Linear, quadratic, and cubic least squares polynomial fit to four points.

Inspection of Fig. 2 shows that the quadratic fit approximates the data points somewhat more closely than the linear fit. The cubic fit, though it passes precisely through each point, is not a significant improvement over the quadratic option—especially since empirical data are often subject to some measurement error. If the empirical data had been taken from the movement of a body in a primary gravitational field with secondary perturbing forces, a quadratic least squares fit would be regarded as the most reasonable smoothing option. On the other hand, if the source of the data had been temperature as a function of altitude, the nature of the problem would indicate that a linear fit is preferable in spite of the fact that the quadratic gives a much better *mathematical* fit. (Perfectionists might argue the premises, but the principle should be clear.) In our example, the quadratic fit would not have approximated the points so well if the fourth point had been somewhat different, for example, $(2, -1/2)$ instead of $(2, -2)$. The principle we emphasize is the best least squares fit is a function of many factors, not the least of which is the visual perception and corresponding judgment of the analyst aided by graphics. Generally, the analyst will make the ultimate decision as to what math model most aptly describes the situation.

3. Mathematical Least Squares (Linear in N Dimensions)

Another common type of least squares fitting is when there is more than one independent variable and it is desired to fit a multidimensional plane. Thus, we would like to fit

$$Y' = A_0 + A_1X_1 + A_2X_2 + \cdots + A_NX_N \qquad (10)$$

in the least squares sense, where the X's are the independent variables and the A's are the coefficients to be derived from the set of data points. Suppose for illustrative purposes we consider two independent variables and use Eq. (10), that is,

$$Y' = A_0 + A_1X_1 + A_2X_2$$

(X_1 and X_2 might be the X and Z coordinates in a three-dimensional fit of a plane. The addition of more variables cannot cause a worse fit than would be the case for any two-dimensional linear fit —vis-a-vis Y' versus X_1 or Y' versus X_2. In general, it would be better.)

The formation equations for the coefficients for the case of two independent variables are

$$A_0 \Sigma 1 + A_1 \Sigma X_1 + A_2 \Sigma X_2 = \Sigma Y$$
$$A_0 \Sigma X_1 + A_1 \Sigma X_1^2 + A_2 \Sigma X_1 X_2 = \Sigma X_1 Y \qquad (11)$$
$$A_0 \Sigma X_2 + A_1 \Sigma X_1 X_2 + A_2 \Sigma X_2^2 = \Sigma X_2 Y$$

(The general formula would be

$$A_0 \Sigma X_N + A_1 \Sigma X_1 X_N + A_2 \Sigma X_2 X_N + \cdots + A_N \Sigma X_N^2 = \Sigma X_N Y \qquad (12)$$

for N independent variables.)

Now consider the same four data coordinates as in the previous example depicted in Fig. 2. The linear fit of that figure showed the least squares line for Y as a function of X. We will extend this to three dimensions by adding another independent variable, Z. In other words, Y will be a function of both X and Z in this case instead of just X. So the point $(X, Y) = (1, 0)$ will be $(X, Y, Z) = (1, 0, 1)$. Similarly, for our example, $(0, 1)$ becomes $(0, 1, 0)$; $(-1, 0)$ becomes $(-1, 0, 1/2)$; and $(2, -2)$ becomes $(2, -2, -1)$. The points are concocted for the example, not derived. Thus, this 3-D example is a special case of Eq. (11); that is, $Y' = A_0 + A_1 X + A_2 Z$ where X and Z represent the X_1 and X_2, respectively. The data are shown in Fig. 3. (The coordinates in the figure are reordered to conform with convention; for example, $(1, 0, 1)$ will be $(1, 1, 0)$ for (X, Z, Y) instead of (X, Y, Z). This is done simply because it is conventional to list the dependent variable last in a coordinate set.)

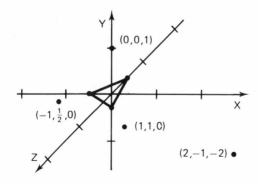

Figure 3   Least squares planar fit to four points.

Now we use Eq. (11) to set up the solutions for the coefficients. We get from the data,

$$\sum_{i=1}^{4} 1 = 4; \qquad \Sigma X_i = 2; \qquad \Sigma Z_i = \frac{1}{2}$$

$$\sum_{i=1}^{4} X_i Z_i = -\frac{3}{2}; \qquad \Sigma X_i^2 = 6; \qquad \Sigma Z_i^2 = \frac{9}{4}$$

$$\sum_{i=1}^{4} Y_i = -1; \qquad \Sigma X_i Y_i = -4; \qquad \Sigma Z_i Y_i = 2$$

Thus,

$$4A_0 + 2A_1 + \frac{1}{2} A_2 = -1$$

$$2A_0 + 6A_1 - \frac{3}{2} A_2 = -4$$

$$A_0 - 3A_1 + \frac{9}{2} A_2 = +4$$

Solving the three derived equations for A_0, A_1, and A_2 gives

$$A_0 = -\frac{1}{4}, \qquad A_1 = -\frac{1}{2}, \qquad A_2 = \frac{4}{7}$$

Thus, the resulting equation is

$$Y' = -\frac{1}{4} - \frac{1}{2} X + \frac{4}{7} Z$$

The traces of this plane in the three principal orthogonal planes are shown in Fig. 3.

Referring back to Eq. (5), we can derive the square of the correlation coefficient for this 3-D least squares fit. We know the values of Y for each (X, Z) because they are given. We find Y' for each (X, Z) from the derived equation above. Recall that $\overline{Y}$ is the average of the given Y's. Therefore, we can determine both $Y - Y'$ and $Y - \overline{Y}$ for each data point, square and sum as indicated in Eq. (5), and complete the determination of R^2. Following this procedure, R^2 for the present 3-D example was derived to be 0.69. For the 2-D linear fit of Fig. 2 where

$$Y' = \frac{1}{10} - \frac{7}{10} X$$

R^2 was derived to be 0.52. We stated earlier that the addition of another variable could not lessen R^2 but would generally increase it.

For our examples, R^2 was increased from 0.52 to 0.69 when we added Z. (The reader might be interested to note that the author found an error in his calculations when it appeared that the 3-D case gave a lower value of R^2 than did the 2-D case. Thus, understanding the theory can be very useful.)

It is instructive to note that we can determine R^2 for Y versus Z as well as Y versus X in the 2-D mode. This turns out to be approximately 0.38. Each 2-D relationship gives its own correlation coefficient, R, which is known as a *simple* correlation coefficient. This is in comparison to the *multiple* correlation coefficient, which is determined from the case of multiple variables. The value of 0.69, which we derived for Y versus both X and Z, is the square of a multiple correlation coefficient. (R would be $\sqrt{0.69} = 0.82$.)

4. Least Squares, Two Interdependent Variables

Graphics can play and is playing a role in data analysis in general and statistical analysis in particular. There is no question that, in many fields, analysts are putting the statistical tools presented here and elsewhere to good use. There is one other type of problem where the least squares and correlation concepts can be aided by graphics—especially in the hand-drawing case in Section A-1. There are many examples where two variables are related, but there is no clear concept of which is dependent or independent, that is, the variables are interdependent. Examples of this are inflation rate and interest rate (each measured as a function of time and each with measurement error), pollution particles in the atmosphere and rainfall over a specified region, and pilot stick position and "g" loading on an airplane during terrain avoidance where unknown wind gusts affect both parameters. In examples of this type, it is more meaningful to perform smoothing by drawing a curve such that the sum of the squares of distances between each point and the curve (perpendicular to the curve) is a minimum. This is in contrast to the more conventional case where one variable is a function of another (Y versus X) and where distances between points and the considered curve are measured vertically. The two concepts are shown in Fig. 4, where curve A represents a conventional least squares fit and curve B represents a curve that comes closest to the points.

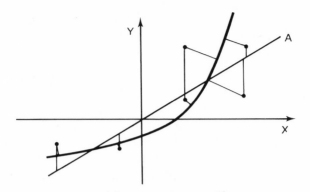

Figure 4 Concept of conventional least squares fitting compared to minimum distance fitting.

In the techniques presented there are some pitfalls, but most requirements for least squares smoothing can be readily met as long as the user has some degree of understanding of what he wants to do. As in all forms of automation, the computer/display system will only make decisions on the nature and type of fit if the decision-making algorithm is provided in advance and is programmed. There is *no* way for the computer to provide the *best* fit or a *smooth* fit to data, automatically, except within a restricted set. Therefore, careful planning based on moderate knowledge of the math tools and options is essential to ensure reliable results.

For least squares, correlation, regression, and so forth, the reader should consult any of a number of standard statistical texts for more detail. However, concepts of fitting by sketching or hand drawing are somewhat specific to interactive graphics and are, therefore, not generally found in the literature.

EXERCISES

1. What is meant by *least squares*?

2. Can a correlation coefficient be determined if a mathematical fit is not used?

3. Which is larger—a simple correlation coefficient or a multiple correlation coefficient?

4. Will a higher degree polynomial fit always give a higher correlation coefficient than that of a lower degree?

5. Why should we not arbitrarily use the highest possible degree polynomial when using least squares polynomial fitting?

6. Determine the least squares line, least squares quadratic fit, and R^2 for each of the following data sets:
  (a) $(-3, -3), (-2, 0), (0, +3), (3, 2), (4, 4)$
  (b) $(-3, -3), (0, 3), (+3, -3)$
  (c) $(-2, 0), (0, 2), (2, 0), (3, -1), (4, -5)$

B. POLYNOMIAL FITTING

As was mentioned in Section A, there are several reasons to use polynomial fittings if and when they can be appropriately applied. Of particular importance is their ease of handling, storing, and computation. It is preferable to perform the multiplications that characterize polynomial computation than it is, for example, to compute square roots or trigonometric functions. The specification of a polynomial to fit data for certain physical problems, such as bridge cables and trajectories, and so forth is quite in order. When a polynomial is specified as a math model to fit data in a least squares sense, certain explicit formulas exist that are comprised of sums of products of the X's and Y's of different powers. This is quite handy.

We may represent a general polynomial of Nth degree by the formula

$$Y = A_N X^N + A_{N-1} X^{N-1} + \cdots + A_1 X + A_0 \qquad (13)$$

where the A's are real (nonimaginary) coefficients. The most commonly used polynomials are when N is unity (linear fit), N is two (quadratic or parabolic fit), and N is three (cubic fit). The linear fit or line can be written in the form

$$Y = A_1 X + A_0$$

and in other convenient forms. The line is a principal requirement in the development of many geometric shapes that are treated later. Also, lines may often have to comply to other constraints, such as tangency to circles and higher degree polynomials. The need for line development to meet constraints occurs in many applications. One of the most common is numerical control part programming. For this application and others, the development of lines is moderately well treated in the literature—for example, APT literature. For this reason, line development is not treated in this section. However, for those who may need additional text on the subject, line development to meet constraints is given in Appendix F.

Discussion of parabolas is given in Section G (Conics), but to characterize the nature of polynomials we will use the cubic, which may be written as

$$Y = A_3X^3 + A_2X^2 + A_1X + A_0 \qquad (14)$$

If information or constraints are sufficiently defined, we can derive the A's to obtain an explicit cubic representation of the data or of the desired design curve. Four conditions are required and that means four points—three points and a first derivative at any one of the three points, two points and two first derivatives at the two points, two points and both first and second derivatives at either of the two points, or one point and three derivatives at the point. In any of these options, the four conditions lead to four linear equations in four unknowns—the four A's. Conventional algebraic techniques will yield their solution.

Whereas cubic polynomials have the advantages described, they and other polynomials cannot, in their nontransformed state, fit data which have multiple values of Y for any range of X's. They are also very sensitive to the slope of the curve to be fit; that is, if a set of data points or a curved shape to be fit is relatively steep, unwanted humps or loops may appear in the polynomial model. We cannot derive a set of A's as coefficients of a polynomial that has a vertical slope. There are ways to use polynomials and get around these problems. This involves spline and parametric techniques. (Discussion of spline and parametric techniques will follow this section.) With deference to these caveats, polynomials are used in many areas of design and analysis. Users of math models should be aware that the mere existence of four conditions does not necessarily mean that a cubic polynomial will meet the problem requirements. As an example of this, suppose we have two lines as shown in Fig. 5. Suppose further that a designer wishes to create a curve from P_1 to P_2 that will not cross either of the two lines. Also, the curve must be tangent to the two lines. These lines are considered in this example to be design boundaries.

Now, with a knowledge of the coordinates of P_1 and P_2 and of the slopes of the two lines (m_1 and m_2), there is no *mathematical* reason why we cannot use the four constraints to derive a cubic polynomial fit. Using the known coordinates as (X_1, Y_1) and (X_2, Y_2) and the known slopes m_1 and m_2 for P_1 and P_2, respec-

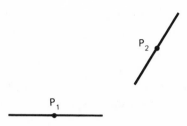

Figure 5 Design constraints.

tively, the four linear equations become

$$Y_1 = A_3X_1^3 + A_2X_1^2 + A_1X_1 + A_0$$
$$Y_2 = A_3X_2^3 + A_2X_2^2 + A_1X_2 + A_0$$
$$Y_1' = m_1 = 3A_3X_1^2 + 2A_2X_2 + A_1$$
$$Y_2' = m_2 = 3A_3X_2^2 + 2A_2X_2 + A_1$$

(15)

These equations can be solved for the A's using any treatment of conventional linear equations (linear or first degree in each A). The resulting cubic equation will pass through P_1 and P_2 and will have the correct slopes, m_1 and m_2. However, there is nothing in the solution process that prevents one or the other of the lines from being crossed. The curve may look like that in Fig. 6, where the dotted portion represents the plot of the cubic beyond the desired P_1 to P_2 range for purposes of visualizing the math shape. This kind of problem can be avoided by use of some other math model, such as certain kinds of elliptic arcs, which will be described in the section on conics.

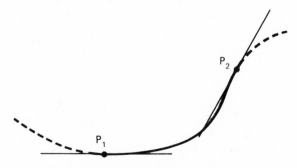

Figure 6 Cubic polynomial crosses a constraint line.

Often polynomials fit other math functions very well. An example of an excellent cubic approximation to some other function is the option to fit a polynomial to a sine function. Suppose we had either an equation of the form $Y = \sin X$ or a set of data points that appear or are known to be from a natural sinusoidal source. Suppose further that we wish to represent this sine function from $-\pi/2$ to $+\pi/2$ radians, a full half cycle. From principles of infinite series in advanced calculus, we may represent $\sin X$ as follows:

$$\sin X = X - \frac{X^3}{3!} + \frac{X^5}{5!} - \cdots \pm \frac{X^N}{N!} \mp \cdots$$

where N is an odd integer that approaches infinity. We may recall that the maximum error for any X is less than the first term omitted in the approximation using a finite number of terms. For example, for small values of X, we may note that X itself is an adequate approximation for $\sin X$. Thus, if X were 0.1, the maximum error for any X between 0 and 0.1 would be $X^3/3! = 0.00017$, which would normally be quite acceptable. In our example, however, we want to cover the full amplitude, and the value of X to reach this maximum Y is $\pi/2$ or 1.57 radians. Using X to approximate $\sin X$ would give a maximum error of $1.57^3/6$ or 65% of $\sin X$, which we will assume to be unacceptable; so we now consider $X - X^3/6$ (the first two terms of the series) to approximate $\sin X$. This yields 0.92 for $\sin \pi/2$, which is, of course, in error by 8%—not too bad. One possibly troublesome problem is that this maximum error occurs at the maximum value of X in the range. In other words, the formula does not give the value $Y = 1$ for $X = \pi/2$, but falls short, giving 0.92 instead. Now, for many practical problems, we want our representative model to fit perfectly at end points of our range of interest, although tiny deviations may be tolerated. We could make this error to be less than any designated tolerance if we simply use more terms from the series. If $X^5/5!$ were added, the approximation would be 0.9996, an error of 0.04%. We see how fast the series converges. Let us return to the fundamentals of cubic polynomials. Consider again $Y = \sin X$, which has slope equal to 1 at $X = 0$ and passes through the three points $(-\pi/2, -1)$, $(0, 0)$, and $(+\pi/2, +1)$. This ensures absolutely no error at the end points. The four conditions will produce the cubic equation of $Y = X - 0.147X^3$. The maximum error between this approximation of $\sin X$ and of $\sin X$ itself occurs at an X of about 1.25 and is 1.4% of the amplitude, which is not bad. This is generally better than what is obtained from the first two terms of the series, plus it closes on the end points. Put another way, any sine curve can be approximated by $X - 0.147X^3$ over an entire half cycle (which implies full cycle because of symmetry) with a maximum error of 1.4% of the amplitude. This formula will apply as long as simple transformations on both X and Y are made such that X (after transformation) varies anywhere between the limits $-\pi/2$ and $+\pi/2$ (the range may be less than $\pi/2$, of course) and Y varies (after transformation) between -1 and $+1$. Such approximations may be useful for graphics and other media because of reduced computation time and ease of use.

EXERCISES

1. What are some of the advantages in using polynomial curve fitting?

2. Why are high degree polynomials generally inappropriate?

3. How many constraints (points and/or derivatives) must be specified to derive a fourth degree polynomial?

4. What options would an analyst consider if there were more constraints than needed; that is, when fitting a quadratic and there are four constraints?

5. Will the specification of four constraints *necessarily* define a cubic polynomial?

6. Can a polynomial be used to fit a set of data when there are two or more values of Y for a particular value of X? Why? What techniques are available to treat this situation?

7. If a slope is specified at a particular point, can the fitted polynomial cross the tangent line representing that slope?

8. Given the points $(-2, 0)$, $(0, 0)$, $(1, -2)$ and $(2, 1)$, derive the cubic polynomial that passes through them.

9. Set up the equations needed to derive the coefficients of a cubic polynomial that passes through $(-1, -1)$ and $(2, 0)$, and has a first derivative of $+2$ and a second derivative of -1 at $(2, 0)$.

10. Derive the quadratic equation for:
    (a) $(0, 0)$, $(2, 0)$, $(3, 0)$
    (b) $(1, 2)$, $(2, 4)$, $Y' = -1$ at $(2, 4)$
    (c) $(2, 4)$, $Y' = -1$, $Y'' = +2$ at $(2, 4)$

11. Using the cubic polynomial of Exercise 8, set up the equations for the coefficients of another cubic that is tangent to that cubic and passes through $(4, 3)$ with a slope of -1.

C. SPLINES

Many applications result in the output of a series of data points that must be fit with a smooth curve that passes through every point. This situation may arise, for example, from the digitizing of a drawing that had been made utilizing plastic curves or other devices. Such plastic curves are referred to by names such as *French curves, ships curves,* or *splines.* It is necessary to derive a math model that goes through all points in order to perform functions such as interpolation, extrapolation, and integration on the curve that the data represents. Intermediate points derived from the fitted model may be used, for example, for precision plotting or for driving a milling machine, ensuring that acceptable tolerances are maintained. The most common technique for fitting an array of points in this way is the mathematical spline technique, which fits a polynomial of designated degree to successive pairs of coordinates —a different set of polynomial coefficients for each pair, which are derived from using the points and by matching the derivatives. This will be explained further in a moment. First, I would like to quote the definition of a mathematical spline as given in the Rogers and Adams text,[1] "In general the mathematical spline is a piecewise polynomial of degree K with continuity of derivatives of order $K - 1$ at the common joints between segments. Thus, the cubic spline has second order continuity at the joints." This definition is consistent with the mathematical representation of the physical spline (explained in the referenced text). The physical spline is the shape of a piece of elastic plastic or wood that is shaped by the use of simple supports. This is the way the contours of ships were laid out, and this was done in a special attic or loft—hence the oft-used term *lofting* with *ships* curves. This physical concept leads to a cubic spline. The above definition of a mathematical spline is a generalization of this concept, of which a cubic spline is a special case. The cubic spline is the lowest degree sequence of polynomials that permits inflection within a given segment. When it is apparent that the points to be fit are not multivalued functions of X and their trend does not indicate any steep slopes (less than unity is a good criterion), then a standard coordinate system may suffice if the numbers are not too large to cause concern over truncation errors.

[1]David F. Rogers and J. Alan Adams, *Mathematical Elements for Computer Graphics* (New York: McGraw Hill, 1976).

Each polynomial segment may be derived in a fixed or rigid XY coordinate system, as follows:

Consider a sequence of points as in Fig. 7.

The successive cubic equations would be

$$Y_1 = A_{(3,1)}X_1^3 + A_{(2,1)}X_1^2 + A_{(1,1)}X_1 + A_{(0,1)}$$

$$Y_2 = A_{(3,2)}X_2^3 + A_{(2,2)}X_2^2 + A_{(1,2)}X_2 + A_{(0,2)}$$

$$\vdots$$

$$Y_9 = A_{(3,9)}X_9^3 + A_{(2,9)}X_9^2 + A_{(1,9)}X_9 + A_{(0,9)}$$

where the subscripts of the coefficients represent the four coefficients of cubics (3, 2, 1, 0) and the number of the particular segment (1, 2, . . . , 9). For subsequent discussions, this general type of notation is a bit cumbersome; so for the special case of cubic polynomials, we will use A, B, C, and D in lieu of A_3, A_2, A_1, and A_0. Then the single subscript will represent the particular segment. Thus,

$$Y_1 = A_1X^3 + B_1X^2 + C_1X + D_1$$

$$Y_2 = A_2X^3 + B_2X^2 + C_2X + D_2$$

$$\vdots \tag{16}$$

$$Y_9 = A_9X^3 + B_9X^2 + C_9X + D_9$$

For N points, the general equation for the ith segment would be

$$Y_i = A_iX_i^3 + B_iX_i^2 + C_iX_i + D_i$$

where i takes on all values: 1, 2, . . . , $N - 1$.

To get started, we wish to derive a cubic fit from P_1 to P_2 using four input conditions. This can be done in any of several ways

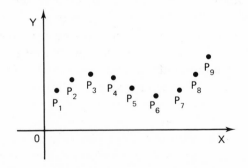

Figure 7 Data points to be fit with a cubic spline.

depending on personal preference and on secondary factors. For example:

1. Use the coordinates of P_1 and P_2 and estimate slopes at both points.
2. Put in a dummy point, P_0. Use its coordinates and the coordinates at P_1 and P_2 and estimate the slope at P_1.
3. Use the first four points, P_1, P_2, P_3, and P_4, to derive an initial cubic. Determine the slopes at P_1 and P_2 from that cubic to derive an adjusted cubic spanning P_1 to P_2.

Etc.

For a cubic spanning the range P_1 to P_2, we then compute Y'' at X_2. This second derivative, the first derivation at X_2, (X_2, Y_2), and (X_3, Y_3) from the next point form the four conditions to derive the adjoining cubic from P_2 to P_3. Succeeding polynomial equations are set up in exactly the same way until $N - 1$ describing equations have been derived to transcend all N points (ten points in our example). This type of fitting could cause computational inaccuracies if the range of coverage of the X's and/or Y's is small compared to their magnitude. Errors that arise from this are caused by the nature of the computation, which leads to truncation errors. Suppose, for example, that ten X values range between 2175.1 and 2177.3. A problem would be avoided if a simple translation is made. Thus, $X^* = X - 2175$ reduces the numbers with which we must deal to a range between 0.1 and 2.3. Such translations are standard practice to minimize truncation errors.

In practice, more often than not, the data or manually-drawn curve to be fit with a mathematical spline develops steep slopes and/or multiple values of Y over some span of X's as in Fig. 8. There are two approaches to this case, each of which has certain advantages and disadvantages. Each concept can be used effectively for a variety of types of curve-fitting problems. One technique is to use transformations coupled with conventional splining procedures. We will designate this technique as *local axis* principles. It has the advantage of keeping slopes relatively small and thus ensuring that there are less wiggles, that the resulting equation relates Y directly to X, that slopes can be matched precisely, and that the procedure is easy to set up. The principle disadvantage is that it is computationally more cumbersome than the other alterna-

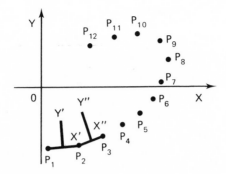

Figure 8 Points to be fit with a spline using a special coordinate treatment.

tive. This alternative is the use of parametric techniques. Such techniques greatly simplify computational time and represent simpler notation. In general, the process does not give precisely continuous derivatives from segment to segment, but this is usually of no practical difficulty. Parametrics can be conveniently extended to N dimensions if required. Slopes in each parametric fit will be less than unity and thus eliminate the concern over steep slopes. On balance, parametric procedures seem to offer more advantages than do local axis procedures. However, both techniques will be discussed in the succeeding sections for the sake of completeness.

EXERCISES

1. Can splines be constructed of consecutive spans of quadratic polynomials?

2. Will a spline pass through each and every point in a spline fit to a set of points?

3. How many sets of polynomial coefficients are necessary to define a spline through ten points?

4. At the joint or connecting point between two spans, what constraints are equated for a cubic spline?

5. Can a span become vertical as a function of X? What procedures are recommended to accommodate vertical slopes or multi-valued functions of X?

6. Develop a spline consisting of three spans through the four points $(-2, -2)$, $(-1, 0)$, $(0, 1)$, and $(2, 0)$.

D. LOCAL AXIS PRINCIPLES

Suppose in Fig. 8, the points P_1, P_2, ..., P_{10} have XY coordinates (X_1, Y_1), (X_2, Y_2), ..., (X_{10}, Y_{10}), respectively. A local origin for an $X'Y'$ coordinate system is chosen at the midpoint between P_1 and P_2 with the X' axis passing through P_1 and P_2 and the Y' axis passing through the midpoint, perpendicular to X'. Let the midpoint for the first segment of the spline be denoted by $(X_{(1,M)}, Y_{(1,M)})$ where

$$X_{(1,M)} = \frac{X_1 + X_2}{2},$$
$$Y_{(1,M)} = \frac{Y_1 + Y_2}{2} \tag{17}$$

Of the many options to determine the first spline segment, we will arbitrarily pick the option where a *dummy* point, P_0, is input (used only for the fitting algorithm, not for subsequent use). Then we wish to use P_0, P_1, P_2, and the slope at P_1 to set up the four constraints to derive the first cubic polynomial. The slope, as well as the points, are recorded in the basic XY coordinate system at this juncture. The slope may be input or may be automatically estimated by having the computer use the slope between points P_0 and P_2 to apply at P_1. (This is not unreasonable as a starting mechanism, especially if P_0 is input approximately with the same ΔX and ΔY between P_0 and P_1 as exists between P_1 and P_2.) We stay away from working with second derivatives because of the cumbersomeness of rotating them as would be required to relate to a new coordinate system. This is no problem with first derivatives as we shall see.

Since we are dealing with P_0, P_1, and P_2, we make a transformation to convert their XY coordinates to $X'Y'$ coordinates. (The reader is asked to note that the "prime" and multiple "prime" notation signifies a change in coordinates. They do not represent derivatives in this treatment. To avoid possible confusion, we will use dY/dX to represent the first derivative.) Thus,

$$X' = \ (X - X_{(1,M)}) \cos \theta_1 + (Y - Y_{(1,M)}) \sin \theta_1$$
$$Y' = -(X - X_{(1,M)}) \sin \theta_1 + (Y - Y_{(1,M)}) \cos \theta_1 \tag{18}$$

where θ_1 is the angle between the X' axis (between P_1 and P_2) and the X axis. The angle θ_1 is, therefore,

$$\theta_1 = \tan^{-1}\left(\frac{Y_2 - Y_1}{X_2 - X_1}\right) \tag{19}$$

The slope in the $X'Y'$ coordinator system may be derived from the slope in the XY system by the following formula:

$$\frac{dY}{dX}\ (\text{in } X'Y') = \tan\left[\tan^{-1}\frac{dY}{dX}\ (\text{in } XY) - \theta_1\right] \tag{20}$$

Now the points and the slope are in $X'Y'$, so the first cubic fit is derived entirely in the $X'Y'$ system. If interpolated data is desired within the segment, it is more convenient to obtain intermediate points in $X'Y'$ and then transform them back to the basic XY system prior to plotting on a plotter or displaying the points on a CRT. Alternatively, the inverse transformation can be performed on the basic equation. In either case, the inverse transformation is

$$\begin{aligned} X &= X' \cos\theta_1 - Y' \sin\theta_1 + X_{(1,M)} \\ Y &= X' \sin\theta_1 + Y' \cos\theta_1 + Y_{(1,M)} \end{aligned} \tag{21}$$

Next an $X''Y''$ system is defined through the midpoint of P_2 and P_3 in the same way as was described for $X'Y'$. The midpoint will be $(X_{(2,M)}, Y_{(2,M)})$ for this second segment and is obtained using P_2 and P_3. The first cubic equation in its $X'Y'$ state is used to get the first derivative at P_2. The points P_1, P_2, P_3, and the derivative at P_2 are transformed in the same way as prescribed for the first segment. In this case, θ_2 is the angle between X'' and X'. The transformed derivative then is

$$\frac{dY}{dX}\ (\text{in } X''Y'') = \tan\left[\tan^{-1}\frac{dY}{dX}\ (\text{in } X'Y') - \theta_2\right] \tag{22}$$

These transformations lead to the second cubic segment which will apply from P_2 to P_3. The procedure to convert data and/or the equation back to the XY system is the same as before. This process is continued until the last point is used. The reader should note that, other than the original data and minimal additional input to initiate the process, the process is recursive and will produce a series of cubic polynomials to spline the entire set of data. The resulting curve will pass through each point without any unwanted inflections.

1. What is the principal advantage of using *local axis* techniques compared to fixed axis principles?

2. Using *local axis* techniques, develop a spline through the following points: (0, 0), (1, 1), (2, 3), (3, 6), (3, 10).

3. There are four data points: $(-2, -2)$, $(0, 0)$, $(1, 4)$, and $(2, 10)$. Derive:
   (a) A conventional cubic through the four points. From that cubic, determine the Y coordinate for $X = 1/2$.
   (b) Place an X' axis through the middle two points with a Y' axis perpendicular at the midpoint of the line joining them. Convert each (XY) coordinate to $X'Y'$. Then derive a cubic of Y' versus X'. Derive Y' for X' values of $-1/2$, 0, and $+1/2$. Then convert each of the resulting interpolated $X'Y'$ coordinates back to the XY system. How do they appear as interpolation values between $(0, 0)$ and $(1, 4)$? How do they compare to interpolation drawn from the basic cubic derived in (a)?

4. Consider a circle with unit radius and center at the origin. Determine the coordinates at:
   (a) $0°$, $10°$, $20°$, $30°$; and
   (b) $75°$, $85°$, $95°$, $105°$.
   Fit a cubic polynomial to each set of four points. Interpolate to get a series of points in the center interval of each polynomial. What are the comparative errors in distance from the origin between the interpolated points and the unit distance on the circle? What is the maximum percent error in the two fits? Does this result suggest anything regarding the value of local axis fitting?

E. PARAMETRIC TECHNIQUES

As discussed in Chap. 1, the basic equation of a circle in XY is

$$(X - h)^2 + (Y - k)^2 = r^2$$

Coordinates for plotting or displaying can be determined by assigning values to X between h and $h + r$ and then solving for Y (when the center and radius are specified). The same equation may be written in several forms—two of which are

$$X = r \cos \theta + h, \qquad Y = r \sin \theta + k$$

and

$$X = r\left(\frac{1 - t^2}{1 + t^2}\right) + h, \qquad Y = r\left(\frac{2t}{1 + t^2}\right) + k$$

(23)

These representations of a circle are called parametric equations. Each of the variables of interest, X and Y, are expressed in terms of a parameter: θ and t for the two examples. By varying the parameter, the variables of concern are generated. This technique is not limited to two dimensions. In fact, the concept can be extended to any number of variables and multiple parameters; for example, in some techniques in surface development (explained later) two parameters are developed into a formula (equation) to represent each of these variables. The advantages of parametric techniques may not be obvious to some users; but they do offer more options to keep computation time down, they represent a better way by which complex relationships may be displayed and visualized, and they represent geometric relationships with simpler math models. Furthermore, certain applications yield parametric relationships in their natural state; motion of an object might be observed in each of three dimensions as a function of time. In such a case, it may be more natural and meaningful to express each variable as a function of time.

When all the information we have is a series of data points as illustrated in Fig. 8, a common and useful practice is to define the parameter as arc length. In this way, no component of arc length (X, Y, or Z in the 3-D case) can exceed the arc length over any range, and hence the first derivative of each component with respect to the parameter will never exceed unity—a desirable attribute as explained in Sections C and D. Thus, no matter how complex the underlying curve or data trend may be, the component curves will be somewhat more bland and easier to treat individually using some basic math model, for example, a spline with a fixed coordinate system. Although knowledge of the exact arc length between data points is desirable, it is not usually known and may be difficult or impossible to obtain precisely. However, if the change in slopes between successive chords is not *too* great, then the chords themselves are usually adequate approximations of incremental arc lengths, at least for purposes of parametric curve fitting. (Maximum error between a chord and the curve to be derived can be established as a function of distance between points and of slopes of adjacent chords as depicted in Fig. 9.) Unless an inflection point exists between P_2 and P_3, the curve must fall within the triangle depicted in Fig. 9. Conventional analytic geometry techniques can be employed to ascertain the height of the triangle from the chord,

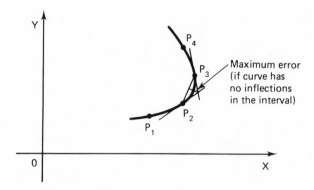

Figure 9 Maximum error from chord to curve.

which is the maximum possible error. Thus, some degree of control can be exercised over the maximum error between the easily-defined chords and the ultimate curve. Sometimes additional data points may be required to comply with specified tolerances.

Now we reflect back to Fig. 7. We will extend the representation of the P_i to be 3-D in nature—adding Z to the X and Y. We define the chords that connect the successive points as $a_1, a_2, \ldots, a_8$. A cubic spline can be derived for each variable X, Y, and Z as a function of a as a parameter. Thus, the recommended technique proceeds as follows. We derive the first cubic segment for each component (variable) independently and similarly. However, we will use only X for illustrative purposes. The starting process can be done in many ways as explained in Section C. We will input the slopes at $a = 0$ and at $a = a_1$ to correspond to points P_1 and P_2. The slopes can be estimated in several ways. A suggested way is to use $(X_2 - X_1)/a_1$ and $(X_3 - X_2)/a_2$ as the two input slopes for X (with similar forms for Y and Z). For gently curving data, this would be a reasonable approach. The four conditions of X's and slopes at P_1 and P_2 suffice to derive the starting or first segment of the spline (in X versus a). Thus,

$$X(a) = A_1 a^3 + B_1 a^2 + C_1 a + D_1 \qquad (24)$$

where $0 \leqslant a \leqslant a_1$.

This technique does not present the same problems as in Section D on local axis principles. There, changing axes caused us to avoid using second derivatives. However, working with arc length as a parameter and holding the axis system constant, we can match any order derivative that we care to across adjoining segments. They will be unaltered in moving across a common point or junction. Therefore, we are at liberty to suggest that X, dY/dX, and d^2Y/dX^2 from the first segment at a_1 be equated to the same functions of the second segment at $a = 0$ (of the second chord where $0 \leqslant a \leqslant a_2$). The second equation is of the form

$$X = A_2a^3 + B_2a^2 + C_2a + D_2$$

Using the common point at P_2, we have

$$X_2 = A_1a_1^3 + B_1a_1^2 + C_1a_1 + D_1 = A_2(0)^3 + B_2(0)^2 + C_2(0) + D_2$$

or simply

$$D_2 = A_1a_1^3 + B_1a_1^2 + C_1a_1 + D_1 \tag{25}$$

Since D_2 is expressed entirely in terms of previously derived data (the coefficients and a_1), D_2 is easily derived. Similarly, setting the first derivatives at P_2 equal gives

$$\frac{dX}{da} = 3A_1a_1^2 + 2B_1a_1 + C_1 = 3A_2(0)^2 + 2B_2(0) + C_2$$

or simply

$$C_2 = 3A_1a_1^2 + 2B_1a_1 + C_1 \tag{26}$$

similarly,

$$\frac{d^2X}{da^2} = 6A_1a_1 + 2B_1 = 6A_2^2(0) + 2B_2$$

or

$$B_2 = 3A_1a_1 + B_1 \tag{27}$$

The fourth coefficient, A_2, is derived by using a_2 at the other end of the curve. Thus,

$$A_2a_2^3 + B_2a_2^2 + C_2a_2 + D_2 = X_3$$

or

$$A_2 = \frac{X_3 - B_2a_2^2 - C_2a_2 - D_2}{a_2^3} \tag{28}$$

Equations (25) through (28) are algebraic expressions which permit us to easily solve for the coefficients of the second segment entirely in terms of previously known or developed data.

These equations may be generalized to give the following

recursive formulas:

$$D_i = A_{i-1}a_{i-1}^3 + B_{i-1}a_{i-1}^2 + C_{i-1}a_{i-1} + D_{i-1} \qquad (29)$$

$$C_i = 3A_{i-1}a_{i-1}^2 + 2B_{i-1}a_{i-1} + C_{i-1} \qquad (30)$$

$$B_i = 3A_{i-1}a_{i-1} + B_{i-1} \qquad (31)$$

$$A_i = \frac{X_{i+1} - B_i a_i^2 - C_i a_i - D_i}{a_i^3} \qquad (32)$$

$$i = 2, 3, \ldots, N - 1 \text{ for an } N \text{ point set}.$$

The element i starts at 2 because a specific starting process was defined for $i = 1$. Thus, the starting process and Eq. (29) through (32) completely define the parametric splining technique for X as well as for each variable that can be expressed as a function of arc length. For those familiar with more concise notation, matrix forms of solution are given in the Rogers and Adams text[2] as is done in other places. Tridiagonal matrices simplify the matrix solutions, but the result is recursive and is not any simpler in comprehension, notation, or execution than what is presented here. Techniques have been developed to use the slopes at each end (at P_i and P_n) as input (whether they are known or estimated). The complete set of polynomials is derived by matrix methods. This process smooths out starting wiggles more quickly.

The preceding procedure describes how X can be generated by successively moving the parameter a from 0 to a_1, 0 to a_2, through 0 to a_{N-1} for the $N - 1$ equations, respectively. This is a bit awkward to keep straight. It is much cleaner to let the total range of a be between zero and some upper limit, a_L. Since it is often desirable to normalize the range of splines to some common base for use in surface fitting, mapping procedures, and so on, it may be useful to make a transformation on each of the spline segments such that a^* has a total range from zero to unity. In such cases, the $N - 1$ segments are each, in turn, generated by a movement of $1/(N - 1)$ of the parameter a^*; that is, a^* varies from 0 to $1/(N - 1)$, from $1/(N - 1)$ to $2/(N - 1)$, through $(N - 2)/(N - 1)$ to $(N - 1)/(N - 1)$ for the successive segments. The transformation to convert each a to a^* for each segment is derived from

$$a = \left[(N - 1)a^* + (1 - i)\right]a_i \qquad (i = 1, 2, \ldots, N - 1) \quad (33)$$

[2]Rogers and Adams, *Mathematical Elements for Computer Graphics.*

The substitution of a in terms of $a*$ in each segment gives, after algebraic simplification, a new series of polynomials that will generate the same X's as before except that the total range of the parameter has been transformed to range from zero unity.

EXERCISES

1. Consider the space curve generated by $(X, Y, Z) = (0, 0, 0)$, $(1, 4, -3)$, $(3, 6, -3)$, $(4, 4, 0)$, $(2, 2, 4)$, and $(1, 4, 6)$.
   (a) Draw a 3-D orthographic system and plot the 3-D points.
   (b) Plot each variable as a function of arc length where arc length is approximated by chord length. (Each plot here is in 2-D.)
   (c) Sketch a curve through each of the three plots.
   (d) Pick an arc length halfway between the second and third points and determine an interpolated X, Y, and Z.

2. A curve passes through the (X, Y) points $(2, 2)$ and $(6, 8)$ with slopes of one and two at the two points, respectively. If there are no inflection points on the curve, determine the maximum error in the interval between the curve and its chord.

3. If a curve passes through $(0, 0)$ with zero slope, determine the maximum slope at $(3, 2)$ to ensure that the maximum error between the curve and its chord between the two points is 0.2.

4. A parametric spline passes through $(1, 2)$, $(4, 6)$, $(9, 18)$ and $(3, 10)$. The equation of the first span for X is

$$X = 0.05a_1^3 + 0.20a_1^2 - 1.65a_1 + 1$$

and for Y is

$$Y = 0.03a_1^3 + 0.10a_1^2 - 0.45a_1 + 2$$

where $0 \leqslant a_1 \leqslant 5$ (chord length of first span).
   Find:
   (a) The range of the parameter a for the succeeding spans.
   (b) The parametric equations of both X and Y versus a_2, a_3, and a_4. (Hint: Each curve varies from zero to chord length.)
   (c) For $a_2 = 3$, solve for X and Y.

5. Using the same starting equation as in Exercise 4, set up the solution for the change in coefficients of the first span for X versus a when the slope at $a_1 = 5$ is changed to $+2$ and all other constraints are unchanged.

6. If a multidimensional function is broken up into each variable as a function of chord length, what is the maximum slope that can exist on any of the functions?

F. BEZIER CURVES

The natural sequence of describing the properties and applicability of polynomials, splines, and parametric techniques can logically be followed by the description of Bezier curves, which have properties of all three previous concepts plus a broader concept of blending or interpolation. The Bezier curves are named after P. E. Bezier of Renault, who is credited with conceiving the procedure (algorithm) by which any span or set of N points may be used to develop a curve with certain special properties. The curve is a polynomial of degree $N - 1$, which may be generated by varying a parameter t between zero and unity. The procedure is applicable to sets of data in any number of dimensions, since each of the component variables is generated independently by the same rule of formation as t varies over its range. Thus, 3-D space curves are just as easy to generate as 2-D curves. The Bezier curves have a number of properties, most of which I believe to be advantageous for many applications—especially applications that are more concerned with aesthetics in design or in styling. Some of the features of Bezier curves with comparison to conventional polynomial or spline fitting are:

1. Bezier curves require no input of derivatives, just data points. In contrast, the requirement to input derivatives for initiating spline fitting is troublesome, since the unknown relationship between curve shape and input derivatives requires subjective decisions for input and thus may lead to some degree of dissatisfaction with initial segments of the fitted spline.
2. Since the degree of a Bezier polynomial depends strictly on the number of data points in a designated span, polynomials of different orders may be automatically developed from span to span. Conventional splines are not very flexible in this regard. The end of one Bezier span, which is the start of the next span, may be made to have common derivatives as explained in detail in the Rogers and Adams text.[3]
3. Bezier curves are particularly well suited to interactive graphics, since the data points are guides or controls to the shape. That is, they influence the curve shape, although the

[3]Rogers and Adams, *Mathematical Elements for Computer Graphics*, p. 142.

curve itself passes through only the first and last points of a designated span of points. The change of location of a data point, done easily at a console, changes the influence on the resulting curve. A graphic user should find it easy to create curve shapes of his preference with relatively little practice and experimentation with how the curve changes with variations in the influencing data points. It should be noted, however, that Bezier curves are inappropriate for applications which necessitate the development of a curve that passes through every point. Bezier curves do not, in general, pass through all points.

4. Since the Bezier technique develops a single equation for each variable, computer storage, I/O, and computation should be somewhat decreased. In contrast, splines require a separate equation for each pair of points. The compactness of the Bezier representation of curves would be quite valuable in approximating masses of data, as in digitizing maps and then fitting the data with a math model.

After the discussion of many of the general characteristics of Bezier curves, the reader may wish to see the definition of the algorithm and examples of its use. There is no loss in generality to specify $N + 1$ points from P_0 through P_N (whereas we used N points from P_1 through P_N in earlier discussions). This is done to simplify the factorial notation which is part of the definition. Thus,

$$P(t) = \sum_{i=0}^{N} P_i \frac{N!}{i!(N - i)!} t^i (1 - t)^{N - i} \tag{34}$$

for $0 \leqslant t \leqslant 1$. (When both i and t are zero, t^i is defined to be unity.)

For any t, the latter part of Eq. (34) is a conventional binomial distribution. Since $t \leqslant 1$, it is equivalent to the probability that is used to weight each P_i. Thus, each P_i has a weight (with unit sum of weight over all P_i) that is unity for P_0 and zero for P_1 through P_N when $t = 0$; less weight is given to P_0 and more weight to succeeding P_i as t increases, reaching a maximum weight for each P_i when t reaches i/N; then weights recede to zero as the weight for P_N becomes unity as t reaches unity. This ensures a shift in the influence of each point as the parameter is moved through its range. Let us take, as an example, four 2-D points as depicted in Fig. 10.

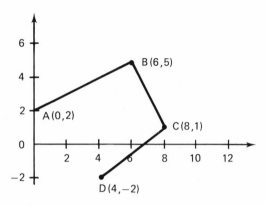

Figure 10 Typical points used to develop a Bezier curve.

From Eq. (34), we note that for each value of t there are four terms to be summed for both X and Y (represented by P), that is,

$$P(t) = (1 - t)^3 P_0 + 3t(1 - t)^2 P_1 + 3t^2(1 - t)P_2 + t^3 P_3 \quad (35)$$

or

$$X(t) = (1 - t)^3 X_0 + 3t(1 - t)^2 X_1 + 3t^2(1 - t)X_2 + t^3 X_3$$

and

$$Y(t) = (1 - t)^3 Y_0 + 3t(1 - t)^2 Y_1 + 3t^2(1 - t)Y_2 + t^3 Y_3$$

where $(X_0, Y_0) \ldots (X_3, Y_3)$ are the four points A, B, C, D, respectively. If we take t in increments of 0.1, the following table results:

t	$X(t)$	$Y(t)$
0	0	2.00
0.1	1.67	2.70
0.2	3.10	3.02
0.3	4.27	3.03
0.4	5.15	2.75
0.5	5.75	2.25
0.6	6.05	1.57
0.7	6.03	0.76
0.8	5.76	−0.14
0.9	5.02	−1.08
1.0	4.00	−2.00

To demonstrate the influencing effect, consider the same four points as in Fig. 10 except for point C, which will be moved

44

horizontally from position (8, 1) to position (14, 1). We will generate a new table of X's for the same increments of t. (Y values will be unaffected because no Y value is changed.)

t	$X(t)$	$Y(t)$
0	0	2.00
0.1	1.84	2.70
0.2	3.68	3.02
0.3	5.41	3.03
0.4	6.88	2.75
0.5	8.00	2.25
0.6	8.64	1.57
0.7	8.71	0.76
0.8	8.00	-0.14
0.9	6.48	-1.08
1.0	4.00	-2.00

By moving point C horizontally, we stretch the X's of the resulting curve without changing the Y's. Figure 11 is now drawn

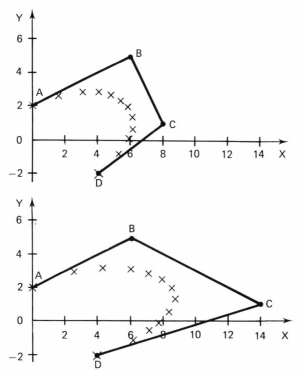

Figure 11 Bezier curves for four points.

to show Fig. 10 with the resulting Bezier curve and the adjusted input with its Bezier curve.

Although the Bezier technique is best suited to "shape" curves and not for passing through multiple points, it can be used to pass through a single point in addition to the end points of a given set. For example, we might desire to move a point from near the right-hand extremity on the first derived curve of Fig. 11 to some specific location and then reconstitute the total Bezier curve.

This would be done by having the computer automatically relocate P_2 without changing any of the other control points of the polygon. The general coefficient of P_2 for the four-point polygon is $3(1 - t)t^2$. We need to know or find the value of t that corresponds to the point on the resulting curve that we would like to be adjusted. The maximum "weight" for a coefficient occurs at $t = i/N$ for P_i. Thus in our example, this occurs at $t = 2/3$ for P_2. We may use any value of t, but the value that gives P_2 the greatest influence is a reasonable choice for this example. For $t = 2/3$, $3(1 - t)t^2$ is $4/9$ or 0.444. Therefore, if we move *only* P_2, the point on the curve that corresponds to $t = 2/3$ will be moved by 44.4% of that for P_2 in both X and Y. We note that $X = 6.07$ and $Y = 1.04$ for $t = 2/3$ (a point generated between that for $t = 0.6$ and $t = 0.7$ as shown in the first table). To illustrate the process, suppose we wish to stretch the Bezier curve that passes through (6.07, 1.04) at $t = 2/3$ such that it will pass through (10, 0) for that t value. Thus, P_2 of the polygon is adjusted from its present position of (8, 1) to (X_a, Y_a) by the following simple process.

$$(X_a - 8)0.444 + 6.07 = 10$$

or

$$X_a = 16.85$$

and

$$(Y_a - 1)0.444 + 1.04 = 0$$

or

$$Y_a = -1.34$$

Thus, the positioning of P_2 (point C) to (16.85, $-$ 1.34) while the other input points remain fixed will cause the resulting curve to pass through point (10, 0). Now the new location of point P_2 and the other three points would be used with $0 \leqslant t \leqslant 1$ to generate an adjusted curve that will have its maximum (in X) close to the (10, 0) location. The change in location of any input point changes the resulting curve; so this Bezier procedure affords the opportunity to create aesthetically pleasing curves by manipulating the input points. Then, if it is desired to make the curve pass through a specific point, an adjustment can be made using an approach similar to that just discussed. One can then pass judgment as to the acceptability of the adjusted curve with the single specified constraint. Multiple specified constraints would be quite cumbersome to satisfy and will not be discussed. The constraint can be processed automatically and the input points can be moved via interactive graphics. The result is the displayed curve that has a math-model description within the computer.

As we said in itemized paragraph 2 of this section, spans of any number of control points can be pieced together quite easily. We may, for example, augment the four points of Fig. 10 with five additional points, $Q_0, Q_1, \ldots, Q_4$, where Q_0 is the same as P_N. If the slope of the line between Q_0 and Q_1 is the same as that between P_{N-1} and P_N, the continuity of slope of the resulting Bezier curves is ensured. The control points, Q, may be depicted along with the P's as in Fig. 12.

A separate Bezier curve of fourth degree (for the five points) can be derived for the set of Q's by reference and analogy to Eqs.

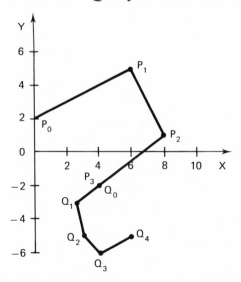

Figure 12 Control points for two successive Bezier curves.

(34) and (35). Thus,

$$Q(t) = (1 - t)^4 Q_0 + 4t(1 - t)^3 Q_1$$
$$+ 6t^2(1 - t)^2 Q_2 + 4t^3(1 - t)Q_3 + t^4 Q_4 \tag{36}$$

As for $P(t)$, Eq. (36) is used for $0 \leqslant t \leqslant 1$ to generate the X's and Y's that correspond to the Q's. Additional spans of any desired number of control points for each may be created virtually at will. Each resulting Bezier curve can be altered within and only within the span that the curve represents. Thus, the relocation of Q_2 would affect the curve from Q_0 to Q_4, would maintain the same end slopes since Q_0, Q_1 and Q_3, Q_4 are not changed, and would leave the first curve, $P(t)$, completely unaffected.

It seems clear that the Bezier concept offers a new and versatile technique to the total repertoire of math models and procedures. The particular method for *weighting* the coordinates has an appeal from the standpoint of reasonableness and practical concern. However, other schemes for weighting the data can be developed, and it is quite conceivable that such alternatives could be preferable for some application types. Thus, not too much mathematical expertise is required for someone to invent his own Bezier-type interpolation. It is obvious, however, that a useful procedure must be predicated on an understanding of the physical nature and characteristics of the applications to which the procedure will be applied.

Although computer programs to develop math models for graphics are not within the scope of this text, the Bezier technique has been selected to illustrate a typical program that has the virtues of ease of implementation coupled with extensive potential utility. Therefore, Appendix G is given to show a sample program for the development of Bezier curves.[4]

Bezier curves are special cases of the considerably more general B-spline curves. In such a system, any degree fit may be specified up to and including that for the Bezier fit. For lower degree fits, the B-spline procedure gives, in general, curves that more closely approach the defining polygon. In fact, second degree fits will define the polygon, per se. Also, the concept of multiple points or knots is a part of the B-spline generalization, which has the effect of pulling the curve closer to the knot. It is the author's

---

[4]This program was provided by Dr. Bertram Herzog, Director, Computing Center, University of Colorado.

opinion that the large number of parameters that must be specified and/or continually changed would be sufficiently complex for most practitioners and would yield so little added value compared to certain available alternatives, that only the most accomplished applied mathematicians should embark on studying B-splines. Comprehension of the physical significance of the parameters and their change is very difficult and probably unnecessary. Because of the above rationale and the fact that this text is intended to be relatively basic, B-splines are not treated here. The Rogers and Adams text[5] contains a comprehensive treatment for the bold of spirit or for those who wish to extend their mathematical base wherever possible.

EXERCISES

1. Are Bezier curves special types of parametric techniques?

2. For a sequence of N points, what degree polynomial as a function of t will result from a Bezier fit?

3. Do successive spans of points have to be fit with the same degree polynomial when using Bezier techniques?

4. Develop the $P(t)$ function for a six-point polygon.

5. For the sequence of three points, (0, 2), (6, 6), and (4, −2), develop $P(t)$ and then plot the Bezier curve using increments of t of 0.1. Now move the third point to (10, 2), and plot the resulting curve in 0.1 increments of t. For each curve, determine the slopes at the end points.

6. Using the equations of this section and the four points (0, 0), (2, 6), (6, 4), and (4, 3) develop the functions of t. Determine how close the resulting curve approaches the second and third points, respectively. Move the second point so that the curve will have a maximum (in the vertical direction) close to $Y = 7$. Now move the third point so that the resulting curve will have a maximum (in X) close to $X = 7$.

7. Suppose we want to use Bezier principles and we wish to have a curve like the basic curve of Exercise 6, except we desire the curve to pass through the midpoint between the second and third points of the set. How can this easily be accomplished *without* changing any of the existing four points? Develop the curve to pass through that midpoint while still using the four basic control points.

8. Can a mix of specific coordinates through which a curve must pass and control points to guide the curve be constituted from Bezier principles? How?

[5]Rogers and Adams, *Mathematical Elements for Computer Graphics*.

9. Develop a five point polygon that closely approximates a circular arc, in the first quadrant, which has unit radius.

10. Develop a four point polygon that approximates $Y = \sin X$ for $0 \leqslant X \leqslant \pi/2$.

G. CONICS

Curves that fall into the conic family are, along with polynomials, the better understood and more frequently used. Physically, they derive their name from the fact that each type (parabola, circle, ellipse, and hyperbola) and degenerate forms of each can be developed by plane cuts at different angles through a cone. The general quadratic equation,

$$AX^2 + BXY + CY^2 + DX + EY + F = 0 \qquad (37)$$

contains all of the conics and their degenerate forms depending on the coefficients. Standard analytic geometry texts explain how to put the general equation into a form to identify and characterize the particular conic.

The conic curves have the quality of smoothness and general aesthetic appeal. As discussed in Section B (Polynomial Fitting), the conics have similar attributes as polynomials in the sense that they are relatively easy and simple to handle from the standpoint of computation, storage, and general compactness. As for polynomials, conics may be specified to fit data or to satisfy a design objective. In this regard, the conic does not have one problem that characterizes polynomials—matching steep slopes. Also, conics do not have inflection. This could be a preferred feature for conics, which was explained in the example of Section B, where the designer wanted to create a curve tangent to two lines but did not want either line to be traversed by the curve. If Eq. (37) is divided by A, the graph of the equation is unaffected. In this form, it should be clear that there are five coefficients that must be derived in order to develop a specific equation and thus define a unique conic. Similar to the technique for polynomial fitting, the specification of a combination of five points and/or derivatives (of various orders) will suffice to solve for the five coefficients B/A, $C/A, \ldots, F/A$ (Eq. (37) divided by A). The resulting equation can be plotted over any prescribed and applicable range of X or Y. The resulting curve will be any one of the conics. Often it is desirable to prescribe a specific type of conic to fit constraints. This arises in our attempt to satisfy design and designer-imposed criteria. We wish to provide the math and computer tools by which the designer may lay out what he currently draws on the drafting board with plastic curves and other mechanical devices—the difference being that it can be done more efficiently and that, as the design is created, its math equivalent is automatically developed and stored in the computer for whatever processes that follow. The intent is to create an optimum balance of functions for person and machine— functions best performed by people are relegated to manual activity while functions best performed by automation are so treated. We will discuss each of the basic conics.

1. Parabolas

The parabolic conic is identical to a second degree or quadratic polynomial and, therefore, the treatment is identical to that of Section B. The basic form is

$$Y = A_0 + A_1X + A_2X^2$$

There must be three constraints to set up the solutions for the coefficients. They may be three points, two points and a slope at one of the points, or one point with both first and second derivatives at that point. Of course, three constraints can lead to other math models unless the quadratic form is explicitly specified as the function to be fit. In meeting design objectives, the parabola is less general than the cubic. However, we can be assured that our curve will have no inflection points if we specify a parabolic fit. The simplicity of form and ease of use in computation make parabolic interpolation quite common. The pitfalls in using parabolas are those explained in Section B for polynomials.

2. Circles

Circles and circular arcs are among the most common geometric forms in application. They are used as parts of symbols and elementary constructions to depict objects, which are, in turn, depicted in schematics and in representative drawings of many types. Typically, circular arcs and straight lines characterize a significant portion of part design. In the aircraft industry, for example, it has been estimated that 80% of the thousands of parts that can be manufactured by numerical control machining are designed entirely by circular arc and line construction. If the designer wishes to create a circle, then three constraints (points, and/or derivatives) are required (aside from the obvious case of providing a center location and a radius). The circle has the special property that both the X^2 and Y^2 coefficients in the general conic equation are equal. Thus, this conic equation may be treated by completing the squares on each variable with the resulting form of Eq. (38),

$$(X - h)^2 + (Y - k)^2 = r^2 \tag{38}$$

where h and k are the coordinates of the center and r is the radius. If the center and radius are given, Eq. (38) gives the precise math model. If the three constraints are given, then the expanded form given by

$$X^2 + Y^2 + aX + bY + c = 0 \tag{39}$$

would normally be more useful in setting up the solutions by substituting points and slopes. The more general conic equation with five coefficients should *not* be used when a circle is prescribed because two of the five constraints would be redundant. Since measured or approximate data may be the source of input, the general equation may give an ellipse with eccentricity nearly zero (almost a circle), but we do not want an "almost" circle. This problem is avoided by using either Eq. (38) or Eq. (39) to develop the circle description. In many practical problems of design, such as numerical control part programming, input of constraints is created indirectly. That is, points and/or slopes may be generated via other natural input conditions, such as tangency to arcs, points at line or arc intersections, and tangency to two lines with a given radius. The natural input criteria lead to input constraints, which, in turn, lead to three linear equations representing the unknown coefficients of Eq. (39). When the circular equation is developed, then it can be displayed or plotted in any of the ways presented in Chap. 1.

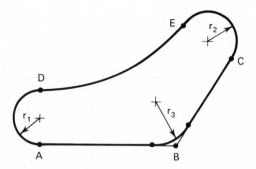

Figure 13 Examples of circular arc constraints and construction.

Examples of circular-arc constraints and construction are depicted in Fig. 13. The process would be as follows. Draw lines from A to B and B to C. Draw a 180° arc of a circle tangent to the line $\overline{AB}$ with radius r_1. Draw a 180° arc of a circle tangent to the line $\overline{BC}$ at C with radius r_2. Draw a circular arc tangent to the two existing arcs at D and E. Draw a final arc tangent to the lines $\overline{AB}$ and $\overline{BC}$ with radius r_3. (The final construction would be to erase the two lines from B to the two points of tangency of the last drawn arc.) A computer graphic construction of a similar bracket showing the addition of some N/C construction is shown in Fig. 14. Figure 15 shows a picture of the finished product—constructed entirely with lines and circular arcs.

Many commercially available graphic software packages have all or most of the common types of circular-arc construction and many other drawing aids. Either this software comes as part of a system package or it may be leased. Although the lease of such software would undoubtedly be significantly less expensive than the development of a comparable package for one's own needs, there are some drawbacks. First of all, the initial monetary outlays for developed packages may not be available for any of a number of reasons. Furthermore, one may not need or even desire a complete package. Generally, it is not feasible to extract portions of a developed package even though only a small portion might be required for certain applications. For these and other reasons, it is desirable for graphic users (as well as users of analytic geometry in general) to comprehend the most concise bases on which arc constructions are predicated. Therefore, the following paragraphs present explicit

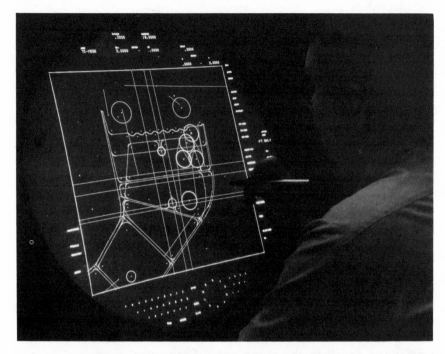

Figure 14 Computer graphic N/C construction.

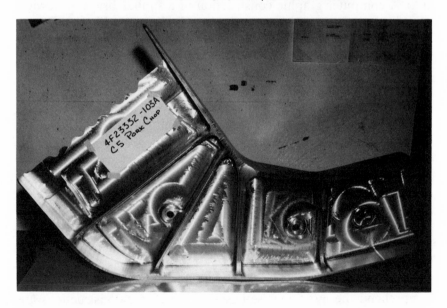

Figure 15 Finished bracket via computer graphic design.

general formulas and certain related special formulas that can readily be used to program individual applications or to perform moderate desk-top calculations. The emphasis is on giving formulas rather than procedures or algorithms, since the latter are generally more involved in both programming and computation time.

a. Circle tangent to two known lines, with a given radius

This case is shown in Fig. 16. For all circular arcs, we must have certain established conventions because there are often several possible solutions (four in this case). Thus, we use a clockwise convention and the position of a graphic tracking symbol with respect to the two lines. In Fig. 16, we wish to put in a clockwise arc from the first position of the tracking symbol, + at l_1, to the second position, + at l_2. There are still two options for the arc location, which are resolved by virtue of the location of the symbols —both above the line in our example. This sets up the construction to locate the center. Thus, suppose that the two lines are given by

$$A_1 X + B_1 Y + C_1 = 0$$

and

$$A_2 X + B_2 Y + C_2 = 0$$

From analytic geometry and the given radius r we can derive the equations of parallel lines at a distance r that will intersect. This solution will be r units from both original lines and therefore is the

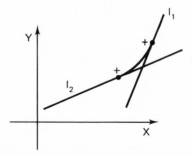

Figure 16 Circle, with a given radius, tangent to two known lines.

center. The two equations (parallel to the given lines) become

$$A_1 X + B_1 Y + C_1 + r \sqrt{A_1^2 + B_1^2} = 0$$

and   (40)

$$A_2 X + B_2 Y + C_2 + r \sqrt{A_2^2 + B_2^2} = 0$$

Equation (40) applies to lines with positive slopes and a tracking symbol above (vertically) the original line. It also applies to an original line with a negative slope when the tracking symbol is below the line. Formula (40) becomes the same with $-r\sqrt{}$ instead of $+r\sqrt{}$ for the opposite conditions—slope positive, symbol below line or slope negative, symbol above line. These are results of principles of analytic geometry and may be used as indicated. The two equations in formula (40) may be solved simultaneously to derive the coordinates of the circle center, (X_C, Y_C). The resulting circular equation from which the desired arc is constructed is

$$(X - X_C)^2 + (Y - Y_C)^2 = r^2 \qquad (41)$$

When only an arc of the total circle is required, some additional level of specification is needed. The initial and terminal points of the arc are determined by determining the intersection of the perpendicular (to the tangents) lines through the appropriate derived circle center with the two tangent lines. With the initial point coordinates and a clockwise convention, successive coordinates are computed until the terminal point is reached. The successive coordinates may be derived easily by translating the center to the origin, by changing the circle to polar form as in Chap. 1, and then by decreasing θ incrementally from its value at the initial point. Decreasing θ will provide clockwise movement from point to point.

As an illustration of case (a) (Section G-2a, this chapter), suppose we have two lines whose equations are

$$l_1, \qquad 2X - Y - 4 = 0$$

and

$$l_2, \qquad X + 2Y - 6 = 0$$

Figure 17 shows the two lines with the desired arc.

In our example, we wish our arc to have a radius of 2 and to be inserted in the clockwise convention as depicted in the figure, that

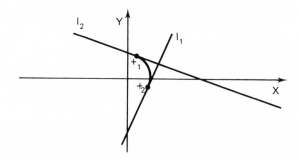

Figure 17 Example of arc construction between two lines.

is, starting at l_2 and terminating at l_1. Applying Eq. (40) and noting that the symbol is below a negatively sloped line at l_2, our equation becomes

$$X + 2Y - 6 + 2\sqrt{1^2 + 2^2} = 0$$

or

$$X + 2Y - 6 + 2\sqrt{5} = 0$$

Similarly, noting that the symbol is above a positively-sloped line in its terminal position, we have

$$2X - Y - 4 + 2\sqrt{2^2 + 1^2} = 0$$

or

$$2X - Y - 4 + 2\sqrt{5} = 0$$

The solution for (X_C, Y_C) is therefore

$$X_C = +0.83 \qquad Y_C = -1.18$$

The resulting circular equation from which the arc is constructed is

$$(X + 0.96)^2 + (Y + 1.03)^2 = 4$$

b. Circle tangent to two circular arcs, with a given radius

Suppose we have two circular arcs whose equations are known. This means that we can find their centers from their equations (if they were not prescribed to initiate the arc definitions). A simple translation is made on both circles, such that the center of one of the circles is at the origin. This makes the derivation and the use of the required formulas much easier. After the center of the desired equation is determined, a simple inverse translation on the center is applied. Figure 18 shows two circular arcs, one of which has its center at the origin.

The two circular equations are

$$X^2 + Y^2 = r_1^2$$

$$(X - a)^2 + (Y - b)^2 = r_2^2$$

and the radius of the connecting arc is to be r_3. The values a and b are the coordinates of the center of the second arc or circle—the one not centered at the origin. In terms of the given data, a, b, r_1, r_2, and r_3, the resulting formulas developed in Appendix A to derive the location of the center (X_C, Y_C) are

$$K_1 = \frac{(r_1 - r_2)(r_1 + r_2 + 2r_3) + a^2 + b^2}{2a}$$

$$K_2 = -\frac{b}{a}$$

$$K_3 = \frac{2K_1 K_2}{K_2^2 + 1}$$

$$K_4 = \frac{K_1^2 - (r_1 + r_3)^2}{K_2^2 + 1} \tag{42}$$

$$K_5 = \frac{-K_3 + (K_3^2 - 4K_4)^{1/2}}{2}$$

$$K_6 = \frac{-K_3 - (K_3^2 - 4K_4)^{1/2}}{2}$$

$$Y_C = K_5 \quad \text{or} \quad K_6$$

$$X_C = K_1 + K_2 K_5 \quad \text{or} \quad K_1 + K_2 K_6$$

There are two mathematical solutions for (X_C, Y_C) as we note in Eq. (42). They come about through the fact that the solutions are the intersection of the common chord with the two given circles that have had each of their radii increased by r_3 (the radius of the connecting arc). However, the single desired solution can be determined simply by computationally noting the location of the tracking symbol when applying the clockwise convention to create the arc from the *first* arc to the *second* arc. That is, the tracking symbol will be closest to the desired solution, and it is a matter of

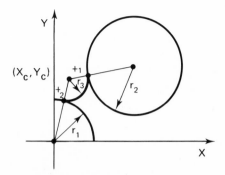

Figure 18 Circular arc tangent to two known circular arcs, with a given radius.

noting the smaller of the distances between K_5 and the symbol compared to K_6 and the symbol. Of course, there are variations on this technique regarding detection of the appropriate solution. However, this discussion should provide the principles and the essential parameters. One caution should be noted. The application of Eq. (42) is predicated on the fact that a solution exists. That is, an arc of specified radius can reach the two existing circles. If no solution exists, then Eq. (42) would produce a negative number under the square root which produces K_5 and K_6. Thus, if $K_2^2 - 4K_4$ is negative, no solution exists. Also, it should be noted that Eq. (42) does not apply, per se, should a be zero and provision should be made accordingly. In this special case, the second given circle will be centered on the Y axis, directly above or below the first. Thus, the parameter a is zero. Referring to the derivation of Eq. (A-5) in Appendix A, we get

$$Y_C = \frac{(r_1 - r_2)(r_1 + r_2 + 2r_3) + b^2}{2b}$$

and

(42a)

$$X_C = \left[(r_1 + r_3)^2 - Y_C^2\right]^{1/2}$$

The preferred X_C is, as described before, ascertained by noting the position of the tracking symbol when the arc definition request is set up. As in case (a) when an arc of the circle is to be drawn, we must find the initial and terminal points of the arc. This is done by first getting the equations of the lines that connect the center of the derived circle with the centers of the given circles. These equations are paired with the respective circle equations and solved to give the desired points. (Additional solutions that are farther from the derived circle center are rejected.) The method of arc generation, per se, was explained in case (a).

As an illustration of the general case, we have two circles with radii 2 and centers $(4, -1)$ and $(2, 3)$. We wish to create an arc of radius 1 that is tangent to the two circles. Figure 19 depicts the example.

We first translate the two circles such that one of them will be centered at the origin; so we decrease the X's by 4 and increase the Y's by 1. The resulting circular equations in the transposed location are

$$X^2 + Y^2 = 2^2$$

and

$$(X + 2)^2 + (Y - 4)^2 = 2^2$$

Thus $r_1 = 2$, $r_2 = 2$, $a = -2$, $b = 4$, and r_3 is given to be 1.

From Eq. (42), we have

$$K_1 = \frac{(r_1 - r_2)(r_1 + r_2 + 2r_3) + a^2 + b^2}{2a}$$

$$K_1 = \frac{0(6) + 4 + 16}{-4} = -5$$

$$K_2 = -\frac{b}{a} = -\frac{4}{-2} = +2$$

$$K_3 = \frac{2K_1K_2}{K_2^2 + 1} = \frac{2(-5)(2)}{4 + 1} = -4$$

$$K_4 = \frac{K_1^2 - (r_1 + r_3)^2}{K_2^2 + 1} = \frac{25 - 3^2}{5} = \frac{16}{5}$$

$$K_5 = \frac{-(-4) + \left[4^2 - 4(16/5)\right]^{1/2}}{2} = 2.9$$

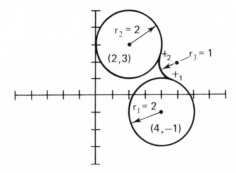

Figure 19 Example of arc tangent to two circles.

It is not necessary to compute K_6, since K_6 must be less than K_5 in the two math solutions for Y_C. It is clear from Fig. 19 that we seek the larger value of Y for the solution of the desired arc. Using K_5, then

$$X_C = K_1 + K_2K_5 = -5 + 2(2.9) = 0.8$$

The translated (X_C', Y_C') is (0.8, 2.9). We merely apply the inverse translation by adding 4 to X_C' and subtracting 1 from Y_C'. Thus, $X_C = 4.8$ and $Y_C = 1.9$. The equation of the circle to draw the arc is

$$(X - 4.8)^2 + (Y - 1.9)^2 = 1^2$$

Since an arc is to be plotted on a graphic display and/or plotter, one final computation is required—the location of the initial and terminal coordinates between which points can be derived using the above equation and techniques described in Chap. 1. For example, the initial point is on the circle whose center is $(4, -1)$. The equation of the line between this center and (X_C, Y_C) above can easily be derived. Its intersection with either of the two appropriate circles gives the initial point plus one other that must be rejected through logical comparisons.

The reader should note that the formulas as presented in cases (a) and (b) are exceedingly easy to use; so that not only should they be quite convenient for programming the computer, but they are quite handy for manual computation with a sliderule or pocket calculator.

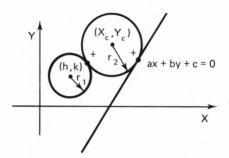

Figure 20 Circle tangent to a given circle and to a given line, with specified radius.

c. Circle tangent to a given circle and a given line with a specified radius.

This case represents, along with the previous two cases, one of the most common requirements. Figure 20 shows the present case. Suppose we have the equation of a circle and the equation of a line. We will translate both equations by h and k (the circle center) such that the translated circle will have its center at $(0, 0)$. As in the previous case, this simplifies the solution equations. We say that the equations, after translation are

$$X^2 + Y^2 = r_1^2$$

and   (43)

$$AX + BY + C = 0$$

The center of the desired circle will be r_2 units from both of the equations in formula (43). Thus we have

$$X^2 + Y^2 = (r_1 + r_2)^2$$

and

$$AX + BY + C = \pm r_2\sqrt{A^2 + B^2}$$

Then

$$X = \frac{\pm r_2\sqrt{A^2 + B^2} - C}{A} - \frac{B}{A}Y$$ (44)

$$X = K_1 + K_2 Y$$

where

$$K_1 = \frac{\pm r_2(A^2 + B^2)^{1/2} - C}{A}$$

$$K_2 = -\frac{B}{A}$$

The positive sign is used for:

1. Positive slope of given line, designated arc (using tracking symbol) below the line.
2. Negative slope, arc above line.

The negative sign is used for the opposite conditions. Substitution and collecting terms give

$$K_3 Y^2 + K_4 Y + K_5 = 0$$

where

$$K_3 = K_2^2 + 1$$
$$K_4 = 2K_1 K_2 \tag{45}$$
$$K_5 = K_1^2 - (r_1 + r_2)^2$$

Then

$$Y_C = \frac{-K_4 \pm \sqrt{K_4^2 - 4K_3 K_5}}{2K_3}$$

This gives two values of Y_C, which give corresponding X_C values when used in Eq. (44). Should $K_3 = 0$, then the solution is single valued and equals

$$-\frac{K_5}{K_4} \tag{45a}$$

The appropriate solution is indicated by the locations of the graphic tracking symbol; that is, using the clockwise convention to determine arc position, the initial (or terminal) tracking symbol will be closer to the preferred circle. This is depicted in Fig. 20 by the + symbol. Had that symbol been placed somewhat lower on the diagram, the other arc-center location would have been indicated as the preference. Again, the arc direction is defined by some convenient convention (we use clockwise). Thus, in Fig. 20 we prefer an arc from the line (initial tracking symbol approximate location) to the circle (terminal location that is shown on the figure).

After the center of the arc has been located, all geometry is translated back again by h and k to depict the relative geometries at their intended locations. Had the given line been vertical (which it would, of course, remain after translation), then it would have the simple form of $X = a$ instead of $AX + BY + C = 0$. Then Y_C is given by

$$Y_C = \pm \left[r_1^2 - (a \pm r_2)^2 \right] \tag{46}$$

The plus inside the parenthesis is used if the circle is to the right of the line and the minus is used if the circle is to the left. (Should the distance from (h, k) to $AX + BY + C = 0$ exceed $r_1 + 2r_2$, then no solution would exist.)

Aside from the special cases (for which provision must be made), Eqs. (45) and (46) cover all cases. The generation of an arc of the resulting circle is similar in principle to that of the first two cases. To illustrate this case, suppose we are given a circle and a line whose equations are $(X - 2)^2 + (Y - 3)^2 = 16$ and $X - Y - 2 = 0$, as shown in Fig. 21. We specify r_2 to be 3. Translating by $(2, 3)$ we have $X^2 + Y^2 = 16$ and $X - Y - 3 = 0$.

According to the rule for K_1, we use the positive form, thus,

$$K_1 = \frac{+3(1^2 + 1^2)^{1/2} - (-3)}{1} = 7.24$$

$$K_2 = +1$$

$$K_3 = 2$$

$$K_4 = 2(7.24)(1) = 14.48$$

$$K_5 = 7.24^2 - (4 + 3)^2 = 3.24$$

The two possible Y_C are -0.24 and -7, but the tracking symbol (its location noted in the computer) was closest to -0.24; so that is the preferred Y_C. Then $X_C = K_1 + K_2 Y_C = 6.996$ (carried to three decimal places). Remembering that we had translated X and Y by -2 and -3, we must reverse the process, which gives

$$(X_C, Y_C) = (8.996, 2.76)$$

Thus the circle equation from which the arc is drawn is

$$(X - 8.996)^2 + (Y - 2.76)^2 = 9$$

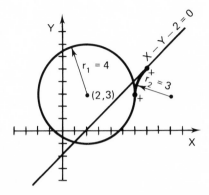

Figure 21 Example of circular arc fit to a circle and a line, with a given radius.

d. Circle tangent to two lines, passing through a given point

This case is shown in Fig. 22.

Inspection of Fig. 22 shows that there are two mathematical solutions. As in previous cases, use of the tracking symbol to designate the two lines and using the clockwise convention, provision can easily be made for the automatic selection of the preferred solution. The solution process is to solve for the location of the center (X_C, Y_C) and then the radius. The distances from both l_1 and l_2 to the unknown solution must be equal in order to satisfy the tangency requirement. Thus, using analytic geometry principles, we derive the equation forms for the two distances and equate them. This results in one equation in X_C and Y_C. The second equation in

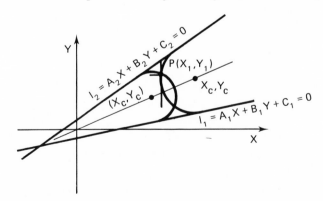

Figure 22 Circle tangent to two lines and passing through a given point.

(X_C, Y_C) is obtained by setting the distance from either of the two lines equal to the distance between (X_C, Y_C) and point, P. After X_C and Y_C are found, r is simply the distance from the center to P. As in previous cases, the solution process and the describing equations are somewhat simpler if initial translations are performed. Therefore, we decrease X by X_1 and Y by Y_1 and assume that

$$A_1X + B_1Y + C_1 = 0$$

and

$$A_2X + B_2Y + C_2 = 0$$

are the equations of the lines after translation. Point P will then be $(0, 0)$. We will use X and Y variables instead of X' and Y' for convenience and then simply add X_1 and Y_1 to the solution to compensate for the original translation. Appendix B contains much of the derivation detail. Thus, we see from that appendix,

$$K_1 = \left(A_1^2 + B_1^2\right)^{1/2}$$

$$K_2 = \left(A_2^2 + B_2^2\right)^{1/2}$$

and

$$(A_1K_2 + A_2K_1)X + (B_1K_2 + B_2K_1)Y + (C_1K_2 + C_2K_1) = 0 \quad (47)$$

or

$$(A_1K_2 - A_2K_1)X + (B_1K_2 - B_2K_1)Y + (C_1K_2 - C_2K_1) = 0 \quad (48)$$

(where Eqs. (47) and (48) are precisely the B-1 and B-2 of Appendix B). Equation (47) or (48) will be used according to the slopes of the lines and the *signs* of the distances, as prescribed in the appendix for a later case of inscribing a circle in a triangle. There is one slight difference here—namely that point P is used instead of a triangle vertex. Thus, the signs of the slopes of l_1 and l_2 and the signs of the distances from l_1 to P and l_2 to P are obtained. In the latter situation, the sign is negative if P is below the line and positive if P is above the line as is determined by comparing the Y coordinates of the line and point for the X coordinate of P. If an odd number of signs is negative (one or three of the four), then Eq. (47) applies. Otherwise, Eq. (48) applies. In either case, we define the coefficients of X, Y, and the constant to be K_3, K_4, and K_5, respectively. Thus, we have

$$K_3X + K_4Y + K_5 = 0$$

or

$$X = -\frac{K_4}{K_3} Y - \frac{K_5}{K_3}$$

We define $-K_4/K_3$ and $-K_5/K_3$ as K_6 and K_7 so,

$$X = K_6 Y + K_7 \tag{49}$$

This is one of the two equations where the solution will be (X_C, Y_C). Now setting the distance from one line to the center equal to the distance from the point to the center, we have

$$\frac{(A_1 X + B_1 Y + C_1)^2}{A_1^2 + B_1^2} = X^2 + Y^2 \tag{50}$$

(since X_1 and Y_1 are zero on the right side). We substitute Eq. (49) into Eq. (50), simplify, and collect terms. This leads to the need to define K_8 and K_9, which are

$$K_8 = A_1 K_6 + B_1 \qquad K_9 = A_1 K_7 + C_1$$

This leads to

$$K_{10} Y^2 + K_{11} Y + K_{12} = 0 \tag{51}$$

where

$$K_{10} = K_8^2 - K_1^2(K_6^2 + 1)$$
$$K_{11} = 2K_8 K_9 - 2K_1^2 K_6 K_7$$
$$K_{12} = K_9^2 - K_1^2 K_7^2$$

It is now merely necessary to use the Y_C portion of Eq. (45) (previous case) to solve for Y_C. In its use, K_{10}, K_{11}, and K_{12} of this case replace K_3, K_4, and K_5 of Eq. (45). Having solved for Y_C, we derive X_C from Eq. (49). (One of the two Y_C solutions is used as previously explained.) Then, r^2 is simply $X_C^2 + Y_C^2$. Finally, we add X_1 to X_C and Y_1 to Y_C to get the actual center (to compensate for initial translations). Thus, we have the desired equation

$$\left[X - (X_C + X_1) \right]^2 + \left[Y - (Y_C + Y_1) \right]^2 = r^2$$

We have shown that from known coefficients (after translation) of A_1, B_1, C_1, A_2, B_2, and C_2 and from input of (X_1, Y_1), we can successively derive K_1 through K_{12} and thereby determine the equation of the circle of concern.

If K_3 is zero, then Y_C is simply equal to $-K_5/K_4$. (K_4 will not be zero when K_3 is zero.) Then X_C would be derived using Eq. (50), setting up the quadratic expression, and solving for X_C using the form of Eq. (45)—as was just done for the general solution for Y_C. If K_{10} is zero, then Y_C is equal to $-K_{12}/K_{11}$, and this solution would be used in Eq. (49) to find X_C. The generation of the appropriate arc is as described in case (a).

We will illustrate the use of these formulas with the following example, which is similar to the diagram of Fig. 22. Suppose the two given lines have equations $X - Y + 1 = 0$ and $X - 3Y - 1 = 0$, and the point that the circular arc must pass through is (4, 3). We first translate these constraints to a local origin such that the point is at (0, 0); that is, $X = X^* + 4$ and $Y = Y^* + 3$. We will rename the adjusted equations in $X^* Y^*$ as XY for convenience—in other words, replace $X^* Y^*$ by another X, Y form. The equations become $X - Y + 2 = 0$ and $X - 3Y - 6 = 0$. Thus, the data to use in the formulas are

$$A_1 = 1, \qquad B_1 = -1, \qquad C_1 = 2$$
$$A_2 = 1, \qquad B_2 = -3, \qquad C_2 = -6$$

then

$$K_1 = \left(A_1^2 + B_1^2\right)^{1/2} = 1.41$$
$$K_2 = \left(A_2^2 + B_2^2\right)^{1/2} = \sqrt{10} = 3.17$$

We use Eq. (47) and not Eq. (48) because the slopes of l_1 and l_2 are both positive and one of the distances is negative, the other positive. Hence, there is an odd number of slope/distance negative signs. Our rule says to use Eq. (47). So,

$$K_3 = A_1 K_2 + A_2 K_1 = 4.58$$
$$K_4 = B_1 K_2 + B_2 K_1 = -7.40$$
$$K_5 = C_1 K_2 + C_2 K_1 = -2.12$$
$$K_6 = \frac{K_4}{K_3} = 1.62$$
$$K_7 = -\frac{K_5}{K_3} = 0.46$$
$$K_8 = A_1 K_6 + B_1 = 0.62$$

$$K_9 = A_1 K_7 + C_1 = 2.46$$

$$K_{10} = K_8^2 - K_1^2(K_6^2 - 1) = -2.86$$

$$K_{11} = 2K_8 K_9 - 2K_1^2 K_6 K_7 = 0.09$$

$$K_{12} = K_9^2 - K_1^2 K_7^2 = 5.63$$

Then we use the quadratic formula as formulated by Eq. (45) where

$$K_{10} Y^2 + K_{11} Y + K_{12} = 0$$

This gives

$$Y_C = 1.43 \quad \text{and} \quad -1.39$$

The corresponding X_C are

$$X_C = 2.78 \quad \text{and} \quad -1.80$$

The preferred solution is determined by the location of the tracking symbol in the initiation process; for example, if the symbol had been placed to the left of $(4, 3)$, then $(-1.80, -1.39)$ would have been indicated (in the local system). However, we will complete both equations. The respective radii would be

$$r^2 = 1.80^2 + 1.39^2 = 5.17$$

and

$$r^2 = 2.78^2 + 1.43^2 = 9.78$$

We still have to add 4 to X_C and 3 to Y_C to compensate for the original translation by those values. The final equations become

$$(X - 2.20)^2 + (Y - 1.71)^2 = 5.17$$

and

$$(X - 6.78)^2 + (Y - 4.43)^2 = 9.78$$

One can see by construction and graphic display that the solutions are reasonable. An automatic checking mechanism can be implemented, if desired, for this and other cases. For example, the distance from the given l_1 to the derived (X_C, Y_C) must equal the distance from the given l_2 to (X_C, Y_C), and the distance is the radius (and this should be true for each (X_C, Y_C) solution).

e. Circle tangent to two circles, passing through a given point

Figure 23 depicts this case. In this case, we have (or can derive from equations) the location of the center and the radius of each of the two circles. Also, we are given—as input via tracking symbol

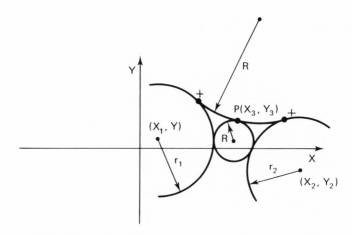

Figure 23 Circle tangent to two circles, and passing through a point.

location, keyboard input, or other medium—the location of a point (X_3, Y_3). In order to derive an arc tangent to the two circles that passes through (X_3, Y_3), we find it expedient for the sake of simplifying our formulas to translate the system by X_3 and Y_3, that is, place the translated origin at the given point. All relative geometry and radii are, of course, unaltered. After deriving the equation of the desired arc and calculating points on the arc for display or plotting, the resulting coordinates are translated back to the original basis prior to the act of displaying or plotting. This case is described, in the translated state, by the three equations that relate the desired center to the known data:

$$(X - X_1)^2 + (Y - Y_1)^2 = (r_1 + R)^2$$
$$(X - X_2)^2 + (Y - Y_2)^2 = (r_2 + R)^2 \qquad (52)$$
$$X^2 + Y^2 = R^2$$

where (X_1, Y_1) and (X_2, Y_2) are the two translated circle centers, r_1 and r_2 are their radii, and R is the radius of the desired circle. The solution of these three equations gives the coordinates of the *mathematical* solutions. If input conditions are unrealistic in the mathematical sense, we will not be able to derive "real" roots. Otherwise, we will derive two solutions that must be resolved by logical means described later. The solution to Eq. (52) for center locations is given in Appendix C and yields the following formulas in terms of given (input) data:

$$K_1 = 2X_2r_1 - 2X_1r_2$$
$$K_2 = 2Y_2r_1 - 2Y_1r_2$$
$$K_3 = r_2(X_1^2 + Y_1^2 - r_1^2) - r_1(X_2^2 + Y_2^2 - r_2^2)$$
$$K_4 = -\frac{K_2}{K_1}$$
$$K_5 = -\frac{K_3}{K_1}$$
$$K_6 = \frac{X_1K_4 - Y_1}{r_1}$$
$$K_7 = \frac{-2X_1K_5 + X_1^2 + Y_1^2 - r_1^2}{2r_1}$$
$$K_8 = K_6^2 - K_4^2 - 1$$
$$K_9 = 2K_6K_7 - 2K_4K_5$$
$$K_{10} = K_7^2 - K_5^2$$

(53)

The ten formulas of Eq. (53) lead to the solution for Y_C when K_8, K_9, and K_{10} are used in the quadratic formula of Eq. (45). Should $K_9^2 - 4K_8K_{10}$ be negative, no solution exists, and the computer program should provide the appropriate output or displayed statement according to personal preferences. For each Y_C solution, $X_C = K_4Y_C + K_5$ and then

$$R^2 = X_C^2 + Y_C^2$$

Should $K_1 = 0$, then Y_C is simply $-K_3/K_2$. (K_2 will not be zero when $K_1 = 0$ for a real situation.) Then we define

$$K_{11} = -\frac{X_1}{r_1}$$
$$K_{12} = \frac{-2Y_1Y_C + X_1^2 + Y_1^2 - r_1^2}{2r_1}$$
$$K_{13} = K_{11}^2 - 1$$
$$K_{14} = 2K_{11}K_{12}$$
$$K_{15} = K_{12}^2 - Y_C^2$$

(54)

Then K_{13}, K_{14}, and K_{15} become the coefficients of the quadratic equation in Eq. (45). In this special case ($K_1 = 0$), we can still have an invalid setup, that is, no solution for X_C if $K_{14}^2 - 4K_{13}K_{15}$ is negative. Due to the nature of the solution process, we could have extraneous roots for X_C only when $K_1 = 0$. The checking process should eliminate them. The distance from (X_C, Y_C) to (X_1, Y_1) minus r_1 should equal the distance from (X_C, Y_C) to (X_2, Y_2) minus r_2, which should equal $X_C^2 + Y_C^2$ (the distance from (X_C, Y_C) to the origin). The one final problem for which provision must be made is which of the two (X_C, Y_C) math solutions is to be used to satisfy the design intent. To do this, each (X_C, Y_C) is treated independently. The equation of the lines jointing (X_C, Y_C) to both (X_1, Y_1) and (X_2, Y_2) are derived and solved simultaneously with $(X - X_1)^2 + (Y - Y_1)^2 = r_1^2$ and $(X - X_2)^2 + (Y - Y_2)^2 = r_2^2$, respectively. These solutions with each circle will contain the points of tangency (plus a set of additional intersections). There will be two pairs of tangencies, one pair for each candidate (X_C, Y_C) as can be seen in Fig. 23. The positioning of a tracking symbol (+ in Fig. 23) at an initial location and then a terminal location will identify which of the two potential (X_C, Y_C) is intended. The symbol locations will be nearer to the tangency points of concern than to the other tangency points. This proximity relationship can be programmed to select the appropriate equation. Then for plotting purposes, this equation and the initial and terminal points are translated (translation within the previous translation) by (X_C, Y_C) such that the desired center is temporarily at the origin. We then define a series of coordinates for plotting by moving from the initial point to the terminal point in our clockwise convention as described in case (a). These coordinates must be increased first by (X_C, Y_C) because of the current translation and then increased by (X_3, Y_3) to compensate for the initial translation. The final plot of the points should fall along the desired arc on the original diagram.

To illustrate this case, suppose we are given two circles with centers at $(1, 2)$ and $(7, 3)$ with radii of 2 and 5, respectively; and suppose we desire an arc to be tangent to the two circles and pass through $(3, -1)$. This is shown in Fig. 24.

The first requirement is translation to put the origin at $(3, -1)$. Thus, (X_1, Y_1) becomes $(-2, 3)$ and (X_2, Y_2) becomes $(4, 4)$ with radii unaffected at 2 and 5, respectively. As stated in the discussion, we will work in this local system. Using these values of

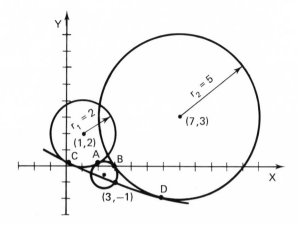

Figure 24 Example of a circle, tangent to two circles, and passing through a point.

X_1, Y_1, X_2, Y_2, r_1 and r_2, we derive from Eq. (53) the following K's, which the reader may verify:

$$K_1 = 36, \qquad K_2 = -14, \qquad K_3 = 31, \qquad K_4 = 0.39,$$

$$K_5 = -0.86, \qquad K_6 = -1.12, \qquad K_7 = 1.39, \qquad K_8 = 0.083,$$

$$K_9 = -2.41, \qquad K_{10} = 1.19$$

Using K_8, K_9, and K_{10} in the quadratic formula, we derive $Y_C = 28.45$ and 0.48. Then from $X_C = K_4 Y_C + K_5$, we get

$$X_C = 10.2 \qquad \text{and} \qquad -0.67$$

For the two solutions, $R = (X_C^2 + Y_C^2)^{1/2}$

$$R = 30.2 \qquad \text{and} \qquad 0.82$$

To check, the distance from (10.2, 28.45) to (−2, 3) is 28.2 plus 2 for r_1 to get to the point of tangency. This totals 30.2 and confirms the radius. Similarly, from (10.2, 28.45) to (4, 4) is 25.2 plus 5 (radius of second circle) which confirms 30.2. Similarly, from (−0.67, 0.48) to (−2, 3), the distance is 2.85 *minus* the r_1, which is calculated to be 0.85 and differs from the computed 0.82 because of round-off. Inspection of Fig. 24 indicates why we must add r in some cases and subtract in others to complete the checking.

It is interesting to observe what happens on implementation of the clockwise convention. Had the initial tracking symbol been near position A and the terminal position near B, we would generate a fillet of less than a 180° arc from A to B, and the generated arc would not pass through the given point, although the complete circle would. Had a position near B been the initial position and near A the terminal position a clockwise convention would generate an arc from B to A through the point—probably not a very practical alternative. Suppose we had indicated a position near C as the initial position and near D as the terminal position. Because of the clockwise convention, our procedure would result in the wrong arc—the large arc going outside of the two given circles. This is because the short arc between the two circles is not clockwise from C to D. It is certainly conceivable that we might think there would be a clockwise fillet from C to D, but such is not the case, and this shows the value of graphics. The visual display of the wrong arc would alert us to this situation. If we truly wanted to generate the short arc of the circle between C and D, we would be obliged to delete the wrong solution and indicate position D as the initial position and C as the terminal position. This would generate the desired arc because of the clockwise programmed convention. This kind of arc is often used for blending purposes. (In the generation of any arc of this discussion, $X_3 = 3$ and $Y_3 = -1$ must be added to all coordinates prior to plotting. The circle center derived to be $(-0.67, 0.48)$ for example, is actually at 2.33 and -0.52. Inspection of Fig. 24 should verify this.)

f. Circle passing through three points

Th❂ situation is rather common and represents one of the easiest circles to derive. Straightforward analytical geometry is applied to derive the center and the radius. Suppose (X_1, Y_1), (X_2, Y_2), and (X_3, Y_3) are the three given points as depicted in Fig. 25. In this situation, the perpendicular bisector of each chord of a circle passes through the center. Since the circle is to pass through each of the three points, the lines joining them are chords. Two of the three perpendicular bisectors will suffice to locate the center. We need only recall the elementary procedures to derive midpoints, determine slopes of lines and their perpendiculars, and derive the equations of lines from the known slopes and points. Without going into the specific details of this process, we would arrive at the following two equations each of which passes through the center (X_C, Y_C). We arbitrarily work with the two chords from the first to the second and the first to the third points. Then the resulting two

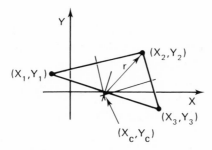

Figure 25 Circle through three points, construction of radius.

equations are

$$(X_2 - X_1)X + (Y_2 - Y_1)Y + \frac{X_1^2 - X_2^2 + Y_1^2 - Y_2^2}{2} = 0$$

and

$$(X_3 - X_1)X + (Y_3 - Y_1)Y + \frac{X_1^2 - X_3^2 + Y_1^2 - Y_3^2}{2} = 0$$

(55)

Now we define

$$K_1 = X_2 - X_1$$
$$K_2 = Y_2 - Y_1$$
$$K_3 = \frac{X_1^2 - X_2^2 + Y_1^2 - Y_2^2}{2}$$
$$K_4 = X_3 - X_1$$
$$K_5 = Y_3 - Y_1$$
$$K_6 = \frac{X_1^2 - X_3^2 + Y_1^2 - Y_3^2}{2}$$

(56)

Thus, Eq. (55) may be rewritten as

$$K_1 X + K_2 Y + K_3 = 0$$
$$K_4 X + K_5 Y + K_6 = 0$$

Solving for Y, we get

$$Y_C = \frac{K_1 K_6 - K_4 K_3}{K_4 K_2 - K_1 K_5}$$

and then

$$X_C = -\frac{K_3 + K_2 Y_C}{K_1}$$

(57)

(If either chord is vertical, formula (57) does not apply, but this special subcase can easily be derived.) Having determined (X_C, Y_C), it is merely necessary to get the distance from any of the points to (X_C, Y_C) to derive the radius.

Now, consider an example where we pick three pairs of coordinates similar to those depicted in Fig. 25, that is, $(-2, 2)$, $(6, 4)$, and $(8, -2)$. Applying Eqs. (56) and (57), we derive

$$K_1 = 6 - (-2) = 8$$

$$K_2 = 4 - 2 = 2$$

$$K_3 = \frac{(-2)^2 - (6)^2 + (2)^2 - (4)^2}{2} = -22$$

$$K_4 = 8 - (-2) = 10$$

$$K_5 = (-2) - (2) = -4$$

$$K_6 = \frac{(-2)^2 - (8)^2 + (2)^2 - (-2)^2}{2} = -30$$

$$Y_C = \frac{8(-30) - (10)(-22)}{(10)(2) - (8)(-4)} = \frac{-20}{52} = \frac{-5}{13} = -0.385$$

$$X_C = -\frac{-22 + 2(-5/13)}{8} = \frac{37}{13} = +2.846$$

The radius, which is measured from (X_C, Y_C) to any of the three points, is calculated to be 5.40. This number can be checked by verifying that the distance from (X_C, Y_C) to each of the other two points is also 5.40. This check is a useful verification of assuring that both the formulas and the arithmetic are correct. Then, the resulting equation is

$$(X - 2.846)^2 + (Y + 0.385)^2 = 5.40^2 = 29.17$$

g. Circle Tangent to Three Lines (Circle Inscribed in a Triangle)

The center of a circle inscribed in a triangle is found by solving for the intersection of two of the three angle bisector lines as depicted in Fig. 26. Either the line equations are known or they can be derived from the given coordinates. The bisector of an angle can be described as a line created by a point moving such that it is

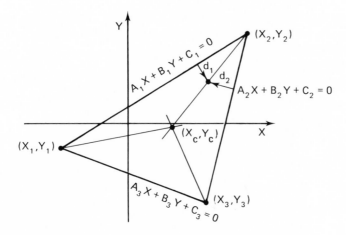

Figure 26   Circle tangent to three lines (construction).

always equidistant from the sides of the angle. That means $d_1 = d_2$ in Fig. 26. The distance from a line to a point is a routine process in analytic geometry. Thus, the distance from $A_1X + BY + C_1 = 0$ to some other (X, Y) point is

$$\frac{A_1X + B_1Y + C_1}{\pm\sqrt{A_1^2 + B_1^2}} = \pm d \tag{58}$$

The sign of the radical is the same as that of the Y coefficient, B. The distance, d, is positive or negative according to whether the point is above or below the line, respectively.

The procedure and formulas to derive the equation for any angular bisector are given in Appendix B. Two such equations suffice to solve for the (X_C, Y_C) coordinates of the circle center. The radius is then found by using Eq. (58), any one of the three line equations, and the specific (X_C, Y_C) point.

Referring to Appendix B, we have for any pair of lines

$$(A_1K_2 + A_2K_1)X + (B_1K_2 + B_2K_1)Y + (C_1K_2 + C_2K_1) = 0 \tag{59}$$

or

$$(A_1K_2 - A_2K_1)X + (B_1K_2 - B_2K_1)Y + (C_1K_2 - C_2K_1) = 0 \tag{60}$$

where

$$K_1 = (A_1^2 + B_1^2)^{1/2} \text{ and } K_2 = (A_2^2 + B_2^2)^{1/2}$$

The subscripts, one and two, are general. However, since there are three lines, the third line will have subscript three. Then the third line can be paired with either of the other two. We arbitrarily pair it with line one. Thus, Eqs. (59) and (60) are identical except that K_2 is replaced by K_3, and A_2, B_2, and C_2 are replaced by A_3, B_3, and C_3, respectively, where $K_3 = (A_3 + B_3)^{1/2}$.

Taking two lines, the rule for selecting either Eq. (59) or (60) is: If the total number of negative signs of the two slopes *and* of the two distances is odd (i.e., one or three), use Eq. (59). Otherwise, use Eq. (60).

The distance sign (not its magnitude) for each line to a point is given by taking the equation, substituting the coordinates of the opposite vertex, and multiplying by ∓ 1 according to whether the slope is $\pm$; that is, the sign of d_1 is derived from the sign of $-(A_1X_3 + B_1Y_3 + C_1)$ if slope $-A_1/B_1$ is positive or from $+(A_1X_3 + B_1Y_3 + C_1)$ if the slope is negative. Similarly, for the second line, the sign of d_2 is taken from $\mp(A_2X_1 + B_2Y_1 + C_1)$ according to the slope $-A_2/B_2$ being $+$ or $-$, respectively. Lines one and three are also paired to give a second angle bisector equation according to Eq. (59) or (60) and the rule of formula selection. The resulting two bisector equations are solved simultaneously to produce the circle center, (X_C, Y_C). Then, as explained in appendix B, the radius, r, is

$$r = \frac{A_1X_C + B_1Y_C + C_1}{\left(A_1^2 + B_1^2\right)^{1/2}} \tag{61}$$

The circle equation becomes

$$(X - X_C)^2 + (Y - Y_C)^2 = r^2 \tag{62}$$

Note: A special case is required should one of the two lines be vertical. Suppose that one angle to be bisected has one side given by

$$A_1X + B_1Y + C_1 = 0$$

and the other side by

$$X + C_2 = 0 \, \text{(vertical line)}$$

Then, the resulting bisecting equation is

$$\left[A_1 + \left(A_1^2 + B_1^2\right)^{1/2}\right]X + B_1Y + \left[C_1 + C_2\left(A_1^2 + B_1^2\right)^{1/2}\right] = 0 \tag{63}$$

Equation 63 is used in lieu of Eq. (59) or (60) when one of the two lines is vertical. There is no problem in using Eq. (59) or (60) should one of the lines be horizontal. No computational problem arises in working, where indicated, with a zero slope.

To illustrate the use of Eqs. (59), (60), and (61) to derive the equation of a circle, suppose the points of Fig. 26 have coordinates

$$(-4, -1.5), (8, 4.5), \text{ and } (5, -5)$$

The three line equations which connect the vertices are determined to be

$$X - 2Y + 1 = 0$$
$$9.5X - 3Y - 62.5 = 0$$
$$3.5X + 9Y + 27.5 = 0$$

Using the coefficients of these equations and the formulas for the K's, we get

$$K_1 = 2.24, \quad K_2 = 9.96, \quad K_3 = 9.66.$$

We must determine whether to use Eq. (59), (60), or (63) to derive two bisecting equations. We note that none of the lines are vertical, so Eq. (63) is rejected. We now refer to the rules by which either Eq. (59) or (60) is selected. Take the first two equations. We note by inspection that their slopes are both positive. We now determine the distance signs, noting that the slopes are both positive. Thus, we have

$$d_1 (\text{sign}) = -(X_3 - 2Y_3 + 1)$$
$$= -[5 - 2(-5) + 1] = -16 = \text{negative}$$
$$d_2 (\text{sign}) = -(9.5X_1 - 3Y_1 - 62.5)$$
$$= -[9.5(-4) - 3(-1.5) - 62.5] = \text{positive}$$

Thus, the two slopes and distances have a total of one negative sign; so we must use Eq. (59) according to the rule. Applying Eq. (59) and using the coefficients from the first two lines, we get the equation of one angle bisector to be

$$31.24X - 26.64Y - 130.04 = 0$$

Now consider the first and third lines. Do we use Eq. (59) or Eq. (60)? We note that the slopes of the first and third equations are positive and negative, respectively. As before, d_1 has a minus sign.

The sign of d_3 is given by

$$+(3.5X_2 + 9Y_2 + 27.5)$$

(since the third line has a negative slope). Thus, we have

$$3.5(8) + 9(4.5) + 27.5$$

which is positive. So the four signs produce two negative signs (not an odd number). Thus, our rule tells us to use Eq. (60) in this instance. The resulting bisecting line has the equation

$$1.82X - 39.48Y - 51.94 = 0$$

to be solved simultaneously with the other bisector,

$$33.24X - 27.28Y - 143.17 = 0$$

Thus, $X_C = 3.16$, $Y_C = -1.17$. Then from use of Eq. (61),

$$r = \frac{3.16 - 2(-1.17) + 1}{(1^2 + 2^2)^{1/2}} = 2.91$$

The equation of the circle is therefore

$$(X - 3.16)^2 + (Y + 1.17)^2 = 2.91^2$$

h. Circle tangent to a line passing through two points

This case is depicted in Fig. 27. There are two possible solutions. The two circle centers lie on the perpendicular bisector of the line connecting the two points, since the connecting line will be a chord of both circles. The centers will be located at positions along the bisector such that, in each case, the distance from either of the points to that position is equal to the distance from the specified line to that position. Thus, d_1 must equal d_2 and d_3 must equal d_4, as depicted in the figure.

In some cases, depending on the slope and the proximity of the tangent line to the two points, the two solutions will lie on the same side of the chord and not between the chord and the line. Given the two input points, P_1 and P_2, we determine the equation of the perpendicular bisector, $A_1X + B_1Y + C_1 = 0$, by conventional means using the midpoint and the slope—the latter being the

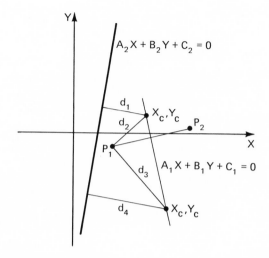

Figure 27 circle tangent to a line (construction) and passing through two points.

negative reciprocal of the slope of the line between P_1 and P_2. The two solutions (for the center) lie on $A_1X + B_1Y + C_1 = 0$. The procedure to find the solutions involves the use of analytical geometry; that is, the normal form of $A_2X + B_2Y + C_2 = 0$ is used to derive the algebraic form of the distance, d, from the line to the unknown solution. The distance from the unknown solution to either P_1 or P_2 is also found in algebraic terms. Then the two distances are equated. This leads to an equation which is a general conic in X and Y. Then X in terms of Y is determined from $A_1X + B_1Y + C_1 = 0$ and substituted into the general conic to yield a quadratic equation in Y. This gives two solutions for Y, each of which gives a corresponding X when substituted into $A_1X + B_1Y + C_1 = 0$. These (X, Y) pairs are the centers of the two solutions. The distances from the centers to either P_1 or P_2 are the radii. On an interactive graphic display, the operator may easily select the preferred solution. Now, as in previous cases, we will present the formulas that apply to this case. Use of the formulas will simplify the solution process. The formulas are as follows: First, derive $A_1X + B_1Y + C_1 = 0$. Compute $(A_2^2 + B_2^2)$ from the specified equation $A_2X + B_2Y + C_2 = 0$. Now compute each of the K's:

$$K_1 = B_2^2$$

$$K_2 = A_2^2$$

$$K_3 = -2A_2B_2$$

$$K_4 = -2\left[X_1(A_2^2 + B_2^2) + A_2C_2\right]$$

$$K_5 = -2\left[Y_1(A_2^2 + B_2^2) + B_2C_2\right]$$

$$K_6 = (A_2^2 + B_2^2)(X_1^2 + Y_1^2) - C_2^2$$

$$K_7 = -\frac{B_1}{A_1}$$

(64)

$$K_8 = -\frac{C_1}{A_1}$$

$$K_9 = K_1K_7^2 + K_3K_7 + K_2$$

$$K_{10} = 2K_1K_7K_8 + K_3K_8 + K_4K_7 + K_5$$

$$K_{11} = K_1K_8^2 + K_4K_8 + K_6$$

This gives

$$K_9Y^2 + K_{10}Y + K_{11} = 0$$

or

(65)

$$Y_C = \frac{-K_{10} \pm \sqrt{K_{10}^2 - 4K_9K_{11}}}{2K_9}$$

If $K_9 = 0$, there is only one solution,

$$Y_C = -\frac{K_{11}}{K_{10}}$$

(66)

Then for each of the two Y_C's (or single Y_C),

$$X_C = K_7Y + K_8$$

$$r^2 = (X_C - X_1)^2 + (Y_C - Y_1)^2$$

Thus,

$$(X - X_C)^2 + (Y - Y_C)^2 = r^2$$

gives the circle equation in each case. For special cases where

$A_1 X + B_1 Y_1 + C_1 = 0$ is either horizontal or vertical, that is, $A_1 = 0$ or $B_1 = 0$, then, K_7 through K_{11} are not computed. The solutions are set up as follows:

1. For a vertical bisector (the two points on a horizontal line), $X = a$, we have the quadratic equation

$$K_2 Y^2 + (aK_3 + K_5)Y + (a^2 K_1 + aK_4 + K_6) = 0$$

For convenience, we let the coefficients of Y and the constant be denoted by K_{12} and K_{13}. Thus,

$$Y_C = \frac{-K_{12} \pm \sqrt{K_{12}^2 - 4K_2 K_{13}}}{2K_2}$$

(which is Eq. (65) with K_2, K_{12}, and K_{13} in place of K_9, K_{10}, and K_{11}). Should A_2 also be zero (which means $K_2 = 0$), then the Y solution is single valued and is simply

$$Y_C = -\frac{a^2 K_1 + aK_4 + K_6}{K_5} \tag{67}$$

(since K_3 is zero when K_2 is zero) and X_C is, or course, equal to a.

2. For a horizontal bisector, $Y = b$, the quadratic equation for the X coordinate of the center is

$$K_1 X^2 + (bK_3 + K_5)X + (b^2 K_2 + bK_4 + K_6) = 0$$

If we denote the X coefficient and the constant by K_{14} and K_{15}, then

$$X_C = \frac{-K_{14} \pm \sqrt{K_{14}^2 - 4K_1 K_{15}}}{2K_1}$$

(which is the same as Eq. (65) with K_1, K_{14}, and K_{15} in place of K_9, K_{10}, and K_{11}). Should B_2 also be zero (which means $K_1 = 0$ and $K_3 = 0$), then the X solution is single valued and is

$$X_C = -\frac{b^2 K_2 + bK_4 + K_6}{K_5} \tag{68}$$

In summary, we have presented here the formulas to derive a circle through two points and tangent to a line for a general case, for the case where the points are on a horizontal line with any tangent line, for the case where the two points are on a horizontal line with a vertical tangent line, for the case where the two points are on a vertical line with any tangent line, and for the case where the two points are on a vertical line and the tangent line is horizontal. These cover all possible situations and are represented by Eqs. (65) through (68). Equation (64) involves the preliminary computations used in one or more of the five cases.

Another apparent special case exists when the tangent line actually passes through one of the two points. (This is equivalent to asking for a circle tangent to a given line at a given point and passing through another point—a common requirement.) However, the formulas presented here for two points and a tangent line will apply to this case; so no special treatment is required.

To illustrate the 2-point, tangent-line case, suppose the two points have coordinates (2, 0) and (4, 2), and the line that the circle will be tangent to is $2X + Y - 16 = 0$. The perpendicular bisector of the chord passes through the midpoint (3, 1) and has a slope of -1 (negative reciprocal of that between P_1 and P_2). Therefore, $A_1X + B_1Y + C_1 = 0$ will be $X + Y - 4 = 0$. From this equation, $A_1 = 1$, $B_1 = 1$, and $C_1 = -4$. From the tangent line, $A_2 = 2$, $B_2 = 1$, and $C_2 = -16$. We note that only Eq. (65) or (66) is applicable (depending on whether K_9 is nonzero or zero), because P_1 to P_2 is neither horizontal nor vertical. Thus, from Eq. (64), we compute the K's.

$$K_1 = 1$$
$$K_2 = 4$$
$$K_3 = -4$$
$$K_4 = -2[2(5) + 2(-16)]$$
$$= 44$$
$$K_5 = -2[0(5) + 1(-16)]$$
$$= 32$$
$$K_6 = 5(4) - 16^2$$
$$= -236$$
$$K_7 = -1$$
$$K_8 = 4$$

$$K_9 = 1(-1)^2 + (-8)(-1) + 4$$
$$= 9$$
$$K_{10} = 2(1)(-1)(4) + (-8)(4)$$
$$+ (44)(-1) + 32$$
$$= -36$$
$$K_{11} = (1)(4^2) + (44)(4) - 236$$
$$= -44$$

Since $K_9 \neq 0$, Eq. (65) applies. Thus,

$$Y_C = \frac{36 \pm \sqrt{36^2 - 4(9)(-44)^{1/2}}}{2(9)}$$

$$Y_C = 4.98 \text{ or } -.98$$

Then from $X + Y - 4 = 0$, which passes through the center,

$$X_C = -.98 \text{ or } 4.98$$

Thus, for $(X_C, Y_C) = (-1.1, 5.1)$,

$$r^2 = (-.98 - 2)^2 + (4.98 - 0)^2 = 33.70$$

Similarly, for $(5.1, -1.1)$

$$r^2 = (4.98 - 2)^2 + (-.98 - 0)^2 = 9.85$$

Thus, the equations for the two solutions are

$$(X + .98)^2 + (Y - 4.98)^2 = 33.70$$

and

$$(X - 4.98)^2 + (Y + .98)^2 = 9.85$$

The input constraints and solutions are depicted in Fig. 28.

At a graphic console, provision can be made by the graphic programmer to reject or accept a particular solution. This might be done by positioning a tracking symbol, which could be put closer to one center than the other—even before the solution, if desired.

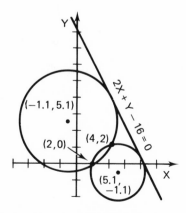

Figure 28 Example of circles tangent to a line and passing through two points.

The reader should note the ease of computation and organization of the solution process. The formulas can also be easily used with a pocket calculator for desk work. This is characteristic of virtually all of the geometric formulations presented in this text.

i. Circle passing through a specified point with a specified slope and second derivative at that point

If the equation of a general circle is taken as

$$(X - X_C)^2 + (Y - Y_C)^2 = r^2,$$

then the derivatives Y' and Y'' are

$$Y' = -\frac{X - X_C}{Y - Y_C}$$

and

$$Y'' = -\frac{(Y - Y_C) - (X - X_C)Y'}{(Y - Y_C)^2}$$

where X, Y, Y', and Y'' are given constraints. Substituting $X - X_C$ in terms of Y' into the third expression and solving for Y_C gives

$$Y_C = \frac{Y'^2 + 1}{Y''} + Y \tag{69}$$

By substituting Y_C into the second expression above and using algebraic simplification, we derive

$$X_C = -\frac{Y'(Y'^2 + 1)}{Y''} + X \tag{70}$$

Then the original expression gives r^2.

As an illustration, suppose the specified point at (2, 1), $Y' = +2$, $Y'' = -1$. From Eq. (69)

$$Y_C = \frac{2^2 + 1}{-1} + 1 = -4$$

From Eq. (70)

$$X_C = \frac{-2(2^2 + 1)}{-1} + 2 = 12$$

Then,

$$(2 - 12)^2 + (1 + 4)^2 = r^2$$

or

$$r^2 = 125$$

Thus we derive

$$(X - 12)^2 + (Y + 4)^2 = 125$$

as the resulting circle equation.

j. Circle tangent to a line, passing through a point, and with a given radius

Suppose the given point has coordinates (X_1, Y_1), the radius is given as r, and the line has equation $AX + BY + C = 0$. The distance from the line to the center is

$$\frac{AX_C + BY_C + C}{\pm (A^2 + B^2)^{1/2}} = \pm r \tag{71}$$

where the sign of the denominator is opposite of the sign of the slope, and $+r$ or $-r$ depends on whether the given point is above or below the line, respectively. Whether it is above or below the line is ascertained by substituting X_1 into $AX + BY + C = 0$, solving for Y, and comparing Y_1 to that solution. We know that (X_1, Y_1) is r units from (X_C, Y_C). Hence,

$$(X_C - X_1)^2 + (Y_C - Y_1)^2 = r^2 \tag{72}$$

Equations (71) and (72) are solved simultaneously to derive X_C and Y_C.

Using Eqs. (71) and (72), we get

$$X_C = \frac{\pm r(A^2 + B^2)^{1/2} - C}{A} - \frac{B}{A} Y_C$$

Let

$$K_1 = \frac{\pm r(A^2 + B^2)^{1/2} - C}{A}$$

where $+$ is used for a positive sloped tangent line with Y_1 below the line or for a negative sloped line where Y_1 is above the line. The minus sign is used for the opposite combinations of slope sign and Y_1 position.

Also, let

$$K_2 = -\frac{B}{A}$$

Then

$$X_C = K_1 + K_2 Y_C$$

Substitution in Eq. (72) and simplification gives

$$K_3 Y_C^2 + K_4 Y_C + K_5 = 0 \tag{73}$$

where

$$K_3 = K_2^2 + 1$$
$$K_4 = 2(K_1 K_2 - X_1 K_2 - Y_1)$$
$$K_5 = K_1^2 - 2X_1 K_1 + X_1^2 + Y_1^2 - r^2$$

The formula for the solution of Eq. (73) is exactly the same as Eqs. (65) and (66), except that the current K_3, K_4, and K_5 replace K_9, K_{10}, and K_{11}, respectively. Then X_C is found by the above formula to be

$$X_C = K_1 + K_2 Y_C$$

for each Y_C. To illustrate this case, we are asked to find the equation of a circle that passes through $(3, -5)$ with a radius of 6 and tangent to $X - 3Y - 4 = 0$, as shown in Fig. 29.

From the K formulas,

$$K_1 = + \frac{+6(1^2 + 3^2)^{1/2} + 4}{1}$$

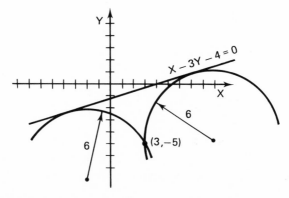

Figure 29 Example of circle tangent to a specified line, through a specified point, and with a given radius.

(based on the sign rule described for K_1). Thus,

$$K_1 = 23.0$$

$$K_2 = -\frac{-3}{1} = +3$$

$$K_3 = 3^2 + 1 = +10$$

$$K_4 = 2[(23)(3) - 3(3) - (-5)]$$

$$= 130$$

$$K_5 = 23^2 - 2(3)(23) + 3^2 + (-5)^2 - 6^2$$

$$= 389$$

Thus,

$$10Y_C^2 + 130Y_C + 389 = 0$$

or

$$Y_C = -4.7, -8.3$$

Then from

$$X_C = K_1 + K_2 Y_C$$
$$X_C = 8.9, -1.9$$

Thus, the two solutions are

$$(X - 8.9)^2 + (Y + 4.7)^2 = 6^2$$

and

$$(X + 1.9)^2 + (Y + 8.3)^2 = 6^2$$

A graphic console operator would then select the preferred solution. If the design situation indicates the need for a circular arc from the point to the line rather than a complete circle, provision would have to be made to initiate the arc at one point and terminate it at another. The plotting technique would be one of those described in Chap. 1. The terminal point may be derived by determining the equation of the line through the center and perpendicular to the tangent line. That equation and the equation of the tangent line are solved simultaneously to get the terminal point. This type of consideration is characteristic of all the circle techniques presented in this section.

The ten circle determination techniques presented in this section do not, of course, include every imaginable situation; but they are the most common, and the set is comprehensive. Formulas are presented that will permit the reader to develop the describing equation without having to be concerned with the analytical procedure or algorithm. The principles of development are presented, however, to afford the reader an opportunity to extend or modify the presented formulas according to particular needs and/or constraints.

EXERCISES

1. Derive the two equations that describe the circular arcs tangent to a line $2X + 3Y - 4 = 0$ and to a line through $(0, 3)$ and $(4, 4)$ with a radius of 3.

2. (a) If two circles have radii of 2 and 3, what is the greatest separation of their centers that will permit a circular arc of radius 1 to be tangent to both?
   (b) If the first has its center at $(1, 1)$ and the second has its center at $(2, 2)$, find the equation of one circular arc that will be tangent to the two circles with radius 6.

3. If a line is the same as that given in Exercise 1, and a circle is the first circle of Exercise 2, find the equation of one of the circular arcs that is tangent to the two with a radius of 2.

4. Given the coordinates $(0, 0)$ and $(2, -4)$, derive the equation of the circle that passes through these points and the intersection of
$$X - Y = 0 \quad and \quad 2X - 5Y - 6 = 0.$$

5. Using the three points of Exercise 4, find:
   (a) The equation of the circle inscribed in the triangle formed by the lines connecting the points.
   (b) The difference in the radii between the circles of Exercises 4 and 5a.

6. Find the equation of a circle that passes through $(-3, -3)$ and $(5, 0)$ and is tangent to
   (a) $X + 3Y - 8 = 0$
   (b) $X + 3Y - 5 = 0$

7. Derive the equation of a circle with the following point, slope, and second derivative constraints:
   (a) $(2, 2)$, -1, -2
   (b) $(0, 0)$, $+2$, -1
   (c) $(2, 2)$, 0, $+1$

8. (a) Find the equation of a circle that passes through $(4, 0)$, is tangent to $X - Y = 0$, and has radius of 10.
   (b) Would there be a solution if the radius were 1.3? Why?

9. Find the equation of a circle tangent to

$$3X - Y + 5 = 0 \quad \text{and} \quad X + 2Y + 4 = 0,$$

which passes through the point (2, 2).

10. Describe the method of sign determination when getting the distance from a line to a point.

3. Ellipses

The requirement to develop an elliptic shape or the math model to describe it is a little more involved than the case for the circle, even though the circle is a particular type of ellipse. In the general conic equation, the ellipse would require that the coefficients of X^2 and Y^2 have the same sign irrespective of the magnitude (in the special case when magnitudes are equal, we have a circle). The ellipse will have both a major axis and a minor axis. When these axes are at an angle with respect to a basic XY coordinate system, then an XY term will exist. On the other hand, when the elliptic axes are parallel to the basic coordinate axes, there will be no XY term in the describing equation. In most cases, we can make a design assumption that the elliptic axes are parallel to the coordinate axes, or we can derive a set of rotated axes in which we would choose to work in order to keep the geometric describing properties in these simpler and more standard terms. Without an XY term, the ellipse is said to be in *standard* position. By completing the squares on the X's and Y's of Eq. (37), the *standard* ellipse may be written as

$$\frac{(X - h)^2}{C_1} + \frac{(Y - k)^2}{C_2} = 1 \tag{74}$$

where h and k are the coordinates of the center and $\sqrt{C_1}$ and $\sqrt{C_2}$ are the semimajor or semiminor (center to vertex) axes. The larger of the two denominators determines whether the major axis is in the X or Y direction. There are four conditions (points and/or derivatives) which must be supplied to solve for the four parameters of Eq. (74). What is not obvious, however, is that some of the four conditions may require some additional restrictions in order to get a solution. This is a subtle fact and such understandings are the backbone of operational interactive design.

a. Standard elliptic arc with horizontal or vertical slope at one end

To illustrate this, consider Fig. 30, where we wish to create an elliptic arc (a portion of a standard ellipse, but not a full quadrant). In our simplified design example, we want to create an elliptic arc between P_1 and P_2 with a horizontal slope at $P_1(m_1 = 0)$ and an input slope m_2, at P_2. We will seek an elliptic arc from an ellipse in standard position whose equation would be in the form of Eq. (74). With this information, the horizontal slope at P_1 implies that the center is somewhere on the Y axis, that is, $h = 0$. The parameters k (location of the center on the Y axis) and C_1 and C_2 must be derived by setting up three equations in three unknowns. This is done, of course, by inputing the XY coordinates of P_1 and P_2 and the desired slope at P_2 into the equation of the derivative. Oddly enough, we cannot input any arbitrary slope and get a solution. It can be shown that m_2 *must* exceed (cannot equal) twice the slope (in magnitude) of the line between P_1 and P_2. This is derived in Appendix D. In mathematical terms,

$$m_2 > 2 \left| \frac{Y_1 - Y_2}{X_1 - X_2} \right|$$

If this is not the case, we would have a computer over-flow—division by zero in the prosecution of the solution. Also, if we do not wish our elliptic arc to have two values (which occurs where the arc is longer than a quadrant), we must not allow our input slope to exceed the vertical position; that is, it must not become positive for this case. (If m_2 were prescribed to be vertical,

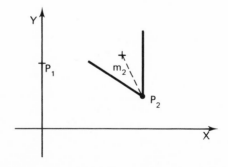

Figure 30   Input conditions for elliptic arc design.

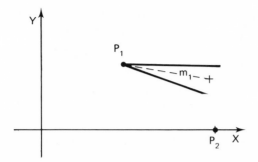

Figure 31 Alternative to Fig. 30 for elliptic arc design.

we would have exactly one elliptic quadrant, k would be Y_2, $\sqrt{C_1}$ would be X_2, and $\sqrt{C_2}$ would be $Y_1 - Y_2$. This is a computationally trivial case.) The value of graphics in aiding the designer can be shown in this situation because a designer may be able to specify that he would like a standard elliptic arc to be drawn between P_1 and P_2. The computer, in turn, can easily compute the tolerance band for the allowable input slope and display the band on a scope. The designer completes the dialogue and the input by placing the tracking symbol between the boundaries as depicted in Fig. 30. The input m_2 will then be the slope of an imaginary line joining the tracking symbol to P_2. The describing equation can then be derived and plotted or displayed. Similar to the example of Fig. 30, a design problem may specify an elliptic arc between P_1 and P_2 with a vertical slope at P_2 as depicted in Fig. 31.

In this case, the input slope at P_1 would have to be between zero (horizontal) and $1/2$ the slope in magnitude of the line between P_1 and P_2. That is,

$$\frac{1}{2} \left| \frac{Y_1 - Y_2}{X_1 - X_2} \right| > m_1 \geqslant 0$$

The region of acceptable slopes must be adhered to if an elliptic arc in standard position with either a horizontal or vertical slope at one end is desired.

Now let us assume that we desire an elliptic equation for an elliptic arc in standard position and that appropriate translations (and/or rotations) have been made such that the center and P_1 are on the Y axis as in Fig. 30. There are three input conditions—$P_1(X_1, Y_1)$, $P_2(X_2, Y_2)$, and the slope at P_2, m. We check m

to see that it exceeds (in magnitude) twice the slope of the line that connects P_1 to P_2 as just described. Using these constraints, we set up three equations and solve for C_1, C_2, and k. This is shown in Appendix E. The resulting equation will be

$$\frac{X^2}{C_1} + \frac{(Y - k)^2}{C_2} = 1 \tag{75}$$

From Appendix E, the successive formulas for k, C_1, and C_2 are

$$k = \frac{Y_2^2 - Y_1^2 - mX_2Y_2}{2Y_2 - 2Y_1 - mX_2} \tag{76}$$

Thus, k is derived entirely from input data. Then

$$C_1 = \frac{X_2(Y_1 - k)^2}{m(k - Y_2)} \tag{77}$$

and finally,

$$C_2 = \frac{C_1 m(k - Y_2)}{X_2} \tag{78}$$

Use of these formulas greatly simplifies the solution process. To illustrate their use, consider a case that will be treated in more detail in subsequent discussions and comparisons. In this case, $P_1 = (0, 6)$, $P_2 = (5, 5)$, and $m = -1$. Using Eq. (76), we derive

$$k = \frac{5^2 - 6^2 - (-1)(5)(5)}{2(5) - 2(6) - (-1)(5)}$$

$$= \frac{14}{3} = 4.67$$

$$C_1 = \frac{5[6 - (14/3)]^2}{-1[(14/3) - 5]}$$

$$= \frac{80}{3} = 26.67$$

$$C_2 = \frac{(80/3)(-1)[(14/3) - 5]}{5}$$

$$= \frac{16}{9} = 1.78$$

Then from Eq. (75) we have

$$\frac{X^2}{26.67} + \frac{(Y - 4.67)^2}{1.78} = 1$$

We see from this example that the use of Eqs. (76) through (78) makes the solution process exceedingly simple. It is certainly better than going through the solution algorithm as was done in Appendix E to derive the formulas.

b. More general, four input condition ellipse

Because of this simplicity, it behooves us to convert, if practicable, any two-point, two-slope situation that requires an elliptic solution to the format just described, that is, one point and the ellipse center on the vertical axis. This means that the general two-point, two-slope case should undergo initial coordinate system transformation to put the given constraints into the desired format, that of case (a) (Section G-3a, this chapter). Figure 32 shows this more general case.

The prescribed procedure is as follows: Put a local $X'Y'$ axis system through the two points. The highest (in Y) of the two points is arbitrarily selected for the Y' axis. The equation of Y' is derived from its slope (negative reciprocal of m_1) and the point $P_1(X_1, Y_1)$. The equation of X' is determined using the slope m_1 and the point $P_2(X_2, Y_2)$. These two equations are solved simultaneously to derive P_3, the $(X\ Y)$ coordinates of the $(X'Y')$ origin. Then the distances from P_3 to P_2 are determined by conventional means. These distances are Y_1' and X_2'; so P_1 in $(X'Y')$ becomes $(0, Y_1')$ and P_2

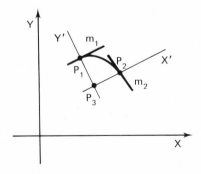

Figure 32 Creating an elliptic arc, given two points and two slopes.

becomes $(X_2', 0)$. These two points and the slope at P_2 in the $X'Y'$ system suffice to use Eqs. (76) through (78) as was described previously. This slope, m_2', is simply derived from given slopes and corresponding angles, that is,

$$m_2' = \tan\left(\tan^{-1}m_2 - \tan^{-1}m_1\right) \tag{79}$$

In Eq. (79), the arc-tan functions are positive or negative acute angles (less than 90° in magnitude) according to whether m is positive or negative, respectively. Of course, the inviolate rule for slopes must be observed to permit a solution, that is, $m_2' > 2$ (slope between P_1 and P_2 in the $X'Y'$ system in magnitude):

$$m_2' > \left| 2\left(\frac{Y_1' - 0}{0 - X_2'} \right) \right|$$

$$m_2' > \left| -2\left(\frac{Y_1'}{X_2'} \right) \right|$$

Also, m_2' must be negative if P_2 is to the right of $P_1,(X_2 > X_1)$ or positive if P_2 is to the left of $P_1,(X_2 < X_1)$. The reader should be cautioned that P_3 is not the center of the elliptic arc nor is the distance P_3 to P_2 the term $\sqrt{C_1}$. P_3 is just a convenient origin for $X'Y'$ when the equation is determined in $X'Y'$. The equation for the ellipse in $X'Y'$ is now determined using Eqs. (76) through (79). Points from P_1 to P_2 can then be computed for ultimate display, on a CRT, or plot, on a plotting device. Each point is rotated and translated to convert it to the basic XY system prior to plotting. Thus, each (X', Y') is rotated by θ as follows:

$$X^* = X'\cos\theta - Y'\sin\theta$$
$$Y^* = X'\sin\theta + Y'\cos\theta \tag{80}$$

where $\theta = \tan^{-1}m_1$. Then, finally

$$X = X^* + X_{P_3}$$
$$Y = Y^* + Y_{P_3}$$

To illustrate the application of Eqs. (76) through (80) and the corresponding tests to derive an elliptic arc from four input constraints, let us suppose that the ellipse is to pass through $P_1(2, 6)$ with a slope of $+1$ and through $P_2(6, 5)$ with a slope of $-3/4$.

Figure 32 is approximately scaled for this case. The two linear equations that are perpendicular and parallel to the first tangent line and which pass through the two points are

$$X + Y - 8 = 0$$

and

$$X - Y - 1 = 0$$

Their solution is P_3, which is $(9/2, 7/2)$—the $X'Y'$ origin.

The distances from P_3 to P_1 and P_2 are calculated to be 3.54 and 2.1, respectively. The slope at P_2 in $X'Y'$ is m_2' as given by Eq. (79), that is,

$$m_2' = \tan\left[\tan^{-1}\left(-\frac{3}{4}\right) - \tan^{-1}(1)\right]$$

$$m_2' = -7$$

Then, the input for Eqs. (76) through (78) is

$$X_1' = 0, \qquad Y_1' = 3.54, \qquad X_2' = 2.1, \qquad Y_2' = 0, \qquad m = -7$$

Using the required slope test, we note that $X_2 > X_1$ and that $|-7| > 2\,|$ slope between P_1 and $P_2|$. This means that we can legitimately proceed to a solution. (Incidentally, for this class of problems, Y_2 is made to be zero by our choice of $X'Y'$ axes. Thus, Eqs. (76) through (78) could be modified by removing terms that contain Y_2 as a factor. However, we will use zero for Y_2 in this illustration.)

From Eq. (76),

$$k = \frac{0 - 12.5 - 0}{0 - 2(3.54) - (-7)(2.1)} = -1.64$$

(k must be negative because we have defined less than an ellipse quadrant and the X' axis goes through P_2.)

From Eq. (77),

$$C_1 = \frac{2.1\left[3.54 - (-1.64)\right]^2}{(-7)(-1.64)} = 5.0$$

From Eq. (78),

$$C_2 = \frac{5(-7)(-1.64)}{2.1} = 27$$

(In this type of problem, this formula can be used as a check because we know that in the $X'Y'$ system C_2 should simply be $(Y'_1 - k)^2$ or 5.18^2—within round-off error of Eq. (78).) Thus, we have

$$\frac{X'^2}{5} + \frac{(Y' + 1.64)^2}{27} = 1$$

From this equation we can get X', Y' for $0 \leqslant X' \leqslant 2.1$, the range of X' for the desired arc. Then we convert $X'Y'$ back to XY for final plotting or display. We will pick one $X'Y'$ point (from the continuous set) to illustrate this process.

From Eq. (80), $\theta = \tan^{-1} 1 = 45°$. As our arbitrary example, we select $X' = 1$ to find the corresponding Y'. Then from the elliptic equation above, $Y' = 3$ when $X' = 1$. Then

$$X^* = X'\cos\theta - Y'\sin\theta$$

$$Y^* = X'\sin\theta + Y'\cos\theta$$

That is,

$$X^* = 1(0.707) - 3(0.707) = -1.4$$

$$Y^* = 1(0.707) + 3(0.707) = +2.8$$

Thus,

$$X = -1.4 + \frac{9}{2} = +3.1$$

$$Y = +2.8 + \frac{7}{2} = +6.3$$

This is one point on the elliptic arc joining P_1 and P_2.

The development of elliptic equations as described in this section should suffice for most practical situations. In principle, we could specify an elliptic arc to satisfy Eq. (74) or (37). Equation (74) leads to a general ellipse in standard position (axes horizontal and vertical), which requires four input constraints. The desired arc would be a portion of that ellipse. The solution process is not unyielding but is considerably more cumbersome than for the cases treated thus far. Equation (37) is a completely general conic which requires five input constraints. Setting up the equations to solve for the coefficients is quite straightforward, although we would have to solve five linear equations simultaneously. Furthermore, the solution may not yield an ellipse since other conics, especially hyperbolas, could well result. The techniques we have described, resulting in Eqs. (76) through (80), do permit us to input slopes at particular points. An elliptic arc that fulfills such constraints represents a smooth or pleasing blend between two points, even though one point is constrained by the formulas to be an ellipse apex. In other words, the underlying apex requirement for Eqs. (76) through (80) is not a severe restriction. For the foregoing reasons, the techniques presented here will suffice for most cases where an elliptic arc is preferred or is considered as a viable mathematical option.

c. An elliptic arc fillet

In the same sense that circular fillets are often desired, there is frequently a need to use an elliptic fillet. We cannot develop a set of formulas for various combinations of constraints for ellipses as we did for circles. However, the following procedure for computer graphics is suggested as a means of creating the desired fillet. Referring to Fig. 33, we would place a tracking symbol *on* a curve. This curve might be a line (as in the figure), a conic, a polynomial, or any mathematical form. The computer would compute the derivative at P_1, and this would be the desired slope or tangency at that point. We would locate another point of tangency in a similar fashion at P_3. In addition, we would indicate a location, P_2, and a tangent line at P_2 with the aid of a tracking symbol and appropriate graphic subroutines. Then if we consider the arc from P_1 to P_3, we note that it is composed of two shorter arcs—P_1 to P_2 and P_2 to P_3. We use the methods of case (b) of this section for each subordinate arc, with the slope at P_2 being made horizontal in the $X'Y'$ system described in that case. As before, Eqs. (76) through (80) would be

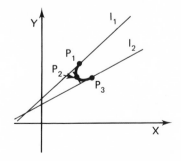

Figure 33 Defining an elliptic fillet.

applicable for each arc and, as such, will produce the desired fillet from P_1 through P_2 to P_3 while maintaining the appropriate tangencies. If the required rule of slopes described in cases (a) and (b) is not met or if the fillet does not have the right aesthetic appeal, we have the freedom to change any or all of the points P_1, P_2, P_3, and the slope at P_2.

d.   Blending of two elliptic arcs

Now we will turn to a typical design situation where certain constraints are defined. Figure 34 shows a box where certain clearances are desired. There are several alternative clearances directed along the X and Y axes, but for each example we specify that the resulting curve and corresponding math model will pass through coordinates (5, 5) with continuity of slope. That is, where two models blend together at (5, 5), their slopes will be required to be equal at that point.

This example differs from the characteristics of those just discussed in that we have points P_1, P_2, and P_3 through which we have an option of blending two arcs or, if it meets all criteria, a single arc. We will attempt to use two *elliptic* arcs in this section. In order to derive two elliptic arcs (top and side), we first have to determine whether solutions exist. That is, can the slope constraints of case (a) be met for both arcs? By setting these constraints equal, we can derive a formula that expresses the threshold X coordinate, X_2, of point P_2 to satisfy the other constraints. Thus,

$$2\left(\frac{Y_1 - Y_2}{X_1 - X_2} \right) = \frac{1}{2} \left(\frac{Y_2 - Y_3}{X_2 - X_3} \right) \tag{81}$$

If we always define an axis system such that P_1 is on the vertical axis and P_3 is on the horizontal axis (and there is no reason we cannot make appropriate axis translations), then $X_1 = 0$ and $Y_3 = 0$. Substitution into Eq. (81) gives the threshold slope equation

$$2\left(\frac{Y_2 - Y_1}{X_2} \right) = \frac{1}{2} \left(\frac{Y_2}{X_2 - X_3} \right) \tag{82}$$

Solving for X_2 gives the threshold or boundary X_2, which we will

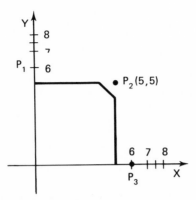

Figure 34 Example of input constraints for conic blending.

now denote as $X_2(B)$. Thus,

$$X_2(B) = \frac{4X_3(Y_2 - Y_1)}{3Y_2 - 4Y_1} \tag{83}$$

It is not at all obvious unless one studies the development very carefully; but Eq. (83) yields the minimum X coordinate for which a slope can be input at P_2 (common to both ellipses), and solutions will exist for a standard elliptic arc from P_1 to P_2 and another from P_2 to P_3. As the first case of our example, suppose $P_1 = (X_1, Y_1) =$ (0, 6) and $P_3 = (X_3, Y_3) = (6, 0)$. ($P_2 = (X_2, Y_2) = (5, 5)$ for all cases.)

If we substitute all data (except X_2) into Eq. (83), we can solve for the $X_2(B)$. In this case, $X_2(B) = 8/3$, which is less than the $X_2 = 5$ that we would like the curves to go through with common slope. Using $X_2 = 5$ (which will give a valid solution) and the other input data in each side of Eq. (82), we see that the slopes must be less than $-2/5$ and greater than $-5/2$, respectively; so if m_c is the common slope, we are at liberty to input any m_c in the range

$$-\frac{2}{5} > m_c > -\frac{5}{2}$$

At a graphic console, that range can be displayed on the screen as in Fig. 35. The tracking symbol may be placed visually between

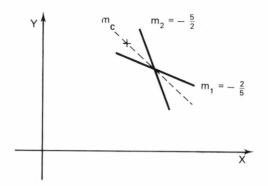

Figure 35 Acceptable range of the input slope, m_c.

the boundary slopes, and then the indicated common slope and the other data will provide the two elliptic arcs. The tracking cross can be moved and new pairs of ellipses derived, if the first set is not acceptable or the console operator wishes to study various alternatives. The only restriction at this juncture is that the tracking symbol that sets up m_C must remain within the boundaries.

As our second case, suppose $P_1 = (0, 7)$ and $P_3 = (7, 0)$. We substitute this data into Eq. (83) and note that the boundary X_2 becomes $X_2(B) = 56/13$, which is still less than 5; so we can still derive two blended elliptic arcs as before. Going to Eq. (82), the range for m_c becomes

$$-\frac{4}{5} > m_c > -\frac{5}{4}$$

This is a more restricted range of options for inputing m_c than in our first case, but it still offers considerable flexibility. If we were to set $P_1 = (0, 5\sqrt{2})$ and $P_3 = (5\sqrt{2}, 0)$, the solution process would lead to a range of ellipses of which a perfect circle from P_1 through P_2 to P_3 is one of the set of acceptable solutions.

As the third case in our example, we use $P_1 = (0, 7.5)$ and $P_3 = (7.5, 0)$. Then $X_2(B)$ becomes precisely 5. Since X_2 must exceed $X_2(B)$, we have reached the limiting position for spacing P_1 and P_3 equidistant from the origin. In other words, if the constraints P_1 and P_3 are equidistant from the origin and less than 7.5, elliptic arcs can be derived for $P_2 = (5, 5)$. (For nonequidistant P_1 and P_3, Eq. (83) would have to be used in each case to derive $X_2(B)$ and compare it to the prescribed X_2.) With this in mind, we will use (0, 8) and (8, 0) for P_1 and P_3 to see what happens. From Eq. (83), $X_2(B)$ is 97/16, which is greater than the prescribed X_2 of 5. Thus, no solutions exist in the conventional elliptic forms we have specified. What might we do, then? This is discussed in the next topic concerning generalized ellipses.

EXERCISES

1. Point P_1 has coordinates $(-2, 3)$ and point P_2 is at $(2, 0)$. If we wish to find the equation of an ellipse in standard position (horizontal slope at P_1):
   (a) What is the range of input slopes at P_2 that will produce a solution?
   (b) What is the ellipse equation for a slope of -2 at P_2?

2. If P_2 of Exercise 1 is to be connected with an elliptic arc to P_3 with coordinates $(3, -4)$ and the slope at P_3 is to be vertical:
   (a) What is the range of input slopes at P_2 that will give a valid solution for an ellipse in standard position?
   (b) What is the range of slopes that is common to both arcs?
   (c) Will the input slope of -2 at P_2 give a solution for the ellipse from P_2 to P_3?

3. It is desired to put a mathematical shape around a box of four by four units, which we can represent by coordinates zero to four in X and Y. We decide to use two joining elliptic arcs with tangency at the point of joining to keep the area between the box and curve at a minimum. To maintain certain clearances, P_1 is assigned the coordinates $(0, 6)$, and P_3 is assigned the coordinates $(8, 0)$. The two arcs from standard ellipses will join along the top of the box at $Y_2 = 4$. What is the closest we can get to the corner of the box?

4. We are given P_1 with coordinates $(2, 4)$ and a slope through P_1 of one. P_2 is given to be $(6, 4)$ with a slope through P_2 of $-3/7$. Find the equation of the standard ellipse that produces an arc from P_1 to P_2.

5. Given two lines with equations $X - Y = 0$ and $2X + Y - 8 = 0$, an elliptic fillet from $(1/2, 7)$ to $(4, 4)$ is desired. Place a point between the two lines with a specified slope to be required there, and set up the equations of two elliptic arcs that pass through the point.

6. How can Eqs. (76) through (78) be employed if an elliptic arc is to be drawn from P_2 with a specified slope at P_2 to P_3, which is on the X axis and has a vertical slope?

4. Generalized Ellipses

Although technically not a conic, it seems appropriate at this point to discuss generalized ellipses[6] in comparison to conventional ellipses. Consider the generalization of the elliptic equation from

$$\frac{X^2}{C_1} + \frac{Y^2}{C_2} = 1 \quad \text{(center at the origin)}$$

to

$$\frac{X^n}{C_1} + \frac{Y^n}{C_2} = 1 \tag{84}$$

Equation (84) permits us to derive one quadrant of a generalized ellipse through *any* set of P_1, P_2, and P_3. When n is greater than unity, this equation will produce a horizontal slope at P_1 and a vertical slope at P_3. When n is unity, we would get a line from P_1 to P_3. When n is less than unity (and greater than zero), the curve will become inverted in the sense that the slope would be vertical at P_1 and horizontal at P_3.

Figure 36 depicts a family of generalized ellipses that are enclosed in a rectangle bounded by P_1 and P_3, and where P_2 is used as a *shoulder* point input along a line from the origin to the corner of the rectangle. (A shoulder point is any intermediate point that helps define the curve shape and to develop its mathematical equation.) The equation for Fig. 36 is

$$\frac{X^n}{P_3^n} + \frac{Y^n}{P_1^n} = 1 \tag{84a}$$

Using formula (84a), we can easily derive the specific equation by inputing P_2 on the diagonal of the rectangle as a fraction of the diagonal's total length, that is, as a fraction of $(P_1^2 + P_3^2)^{1/2}$. This kind of input is quite suitable to computer graphics. We will denote that fraction by the parameter a, where a varies from zero to unity. Using proportionalities, we note that for any a the XY coordinates of the shoulder point that is indicated on the diagonal will be aP_3 and aP_1, respectively. Then from formula (84a),

$$\frac{X^n}{P_3^n} + \frac{Y^n}{P_1^n} = \frac{(aP_3)^n}{P_3^n} + \frac{(aP_1)^n}{P_1^n} = 1$$

[6]Generalized ellipses are often called *super* ellipses.

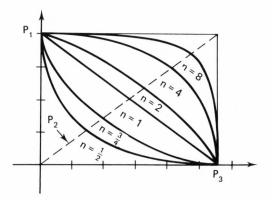

Figure 36 Family of generalized ellipses.

or

$$a^n + a^n = 1$$

$$a^n = \frac{1}{2} \tag{84b}$$

$$n = \frac{\log 1/2}{\log a}$$

This solution for n and the given P_1 and P_3 suffice to satisfy Eq. (84a) and hence Eq. (84).

Returning to the case in Section G-3d, which would not accommodate the matching of two conventional elliptic arcs, we can derive a single math model using the various forms of Eq. (84). In the case of interest, $P_1 = P_3 = 8$ and $P_2 = (5, 5)$. From this, $a = 5/8$. Therefore,

$$n = \frac{\log 1/2}{\log 5/8} = \frac{-0.30103}{-0.20412} = 1.475.$$

Then

$$C_1 = C_2 = 8^n = 21.47.$$

So,

$$X^{1.475} + Y^{1.475} = 21.47$$

is the equation that meets the requirements of passing through (0, 8), (5, 5), and (8, 0) with horizontal slope at the first point and vertical slope at the last.

Even in the first case of Section G-3 where $P_1 = (0, 6)$ and $P_3 = (6, 0)$, we could use the generalized ellipse as an option in lieu of using two blending elliptic arcs. Solving for the parameters as before, the equation becomes

$$X^{3.802} + Y^{3.802} = 908.6$$

Such equations as in these latter examples will define a continuous curve from P_1 to P_3 through any P_2. One may then ask, why do we bother with the two-arc option? The answer is that the two-arc option offers somewhat more flexibility because, in most cases, we have a range or broad selection of input slopes at P_2 to help shape the total curve. No slope option exists when using the generalized ellipse. In the generalized elliptic case, we can derive one and only one curve for a given set of three points. We have to take it or leave it. However, for some sets of three points, the two-arc blending cannot be obtained (within the type functions specified) and, therefore, the generalized ellipse may be a good way out. (It may be helpful to know that the slope of the generalized ellipse is $-P_1/P_3$ where the curve crosses the diagonal. In particular, for $P_1 = P_3$, this slope is always -1.) Whether or not to use the conventional ellipse or the generalized ellipse when both options exist should be a concern for the designer of the system, not necessarily for the user of the system, although user expertise should be helpful in the original system definition. The use of different exponents for X and Y adds flexibility but also adds complexity. The additional generality is not considered to be of sufficiently broad utility to be given special treatment in this text.

We have shown that the general ellipse is a single-function alternative to the blending of two elliptic arcs, although we lose the freedom of having a range of possible input slopes at the intermediate point (the point we identify as P_2). Where we employed a generalized ellipse to pass through $(0, 6)$, $(5, 5)$, and $(6, 0)$, we derived $X^{3.802} + Y^{3.802} = 908.6$. The slope of this curve at $(5, 5)$ is -1. It should add insight to the comparison of different curve-fitting options if we compare the general ellipse for $(0, 6)$ to $(5, 5)$ with the standard ellipse over the same range where we specify the same slope at $(5, 5)$, that is, -1. While making this comparison, we will add a comparison with a Bezier curve over the same range and with the same slope at $(5, 5)$. To affect the horizontal slope at $(0, 6)$ and the -1 slope at $(5, 5)$ for the Bezier curve, we use the control points (according to Bezier principles) of (0, 6), (4, 6), and (5, 5). The equations that define the three curves are

$$X^{3.802} + Y^{3.802} = 908.6 \quad \text{for the general ellipse}$$

$$\frac{X^2}{20.8} + \frac{(Y - 4.33)^2}{2.78} = 1 \quad \text{for the standard ellipse}$$

$$P = (1 - t)^2 P_0 + 2t(1 - t)P_1 + t^2 P_2 \quad \text{for the Bezier curve}$$

where P_0, P_1, and P_2 are 0, 4, and 5 when solving for X, and 6, 6, and 5 when solving for Y.

We wish to plot Y as a function of X for integral values of X, that is, $X = 1, 2, 3,$ and 4. In the Bezier case we first solve for the t that corresponds to the desired X's. Thus we would have $t = 0.1313, 0.2790, 0.4515,$ and 0.6667 for $X = 1, 2, 3,$ and 4, respectively.

The comparison of curves for the three methods is shown in the following table:

X	General Ellipse	Standard Ellipse	Bezier Curve
0	6.000	6.000	6.000
1	5.996	5.975	5.975
2	5.974	5.896	5.922
3	5.873	5.762	5.796
4	5.630	5.511	5.556
5	5.000	5.000	5.000

Inspection of the table indicates that there is very little difference, although the Bezier is between the other two and closer to the standard ellipse. The objective of making the comparison is to show that for many examples, personal preferences or the existence of a particular technique may have more of an effect (with justification) on the type of function to use than other considerations.

When speaking of design and when going through the various math model options, the reader may get off-track by thinking that the designer *must* be concerned with the math. That is not the case at all, or it should not be the case for most applications. Whatever equations are set up, they are set up *automatically* and solved *automatically*. This is the province of the systems programmers and analysts who must provide the design tools. The designer at a console need only specify which of the available options he intends to use and then specify the constraints such as input points and/or slopes. He should understand the relative suitability of the available options. The computer will, hopefully, do the rest as programmed. This does not mean that knowledge of the underlying math functions would not be helpful to the user when the original specifications are set up to solve his problem.

EXERCISES

1. Given the points $(0, 4)$ and $(4, 0)$, find the equation of the general ellipse that passes through these points and $(3, 3)$.

2. Given $X^4/81 + Y^4/16 = 1$:
   (a) What is Y for $X = 2$?
   (b) What is the slope at that point?

3. (a) How does the curve of Exercise 2 compare with $X^2/81 + Y^2/16 = 1$ at $X = 1$ and $X = 2$?
   (b) What are the X and Y intercepts of the two equations?

4. How many constraints are necessary to derive a generalized ellipse?

5. Using the principles of calculus, verify the text formula for the slope where the generalized ellipse crosses the diagonal of the rectangle. Hint: use aP_3 and aP_1 as the diagonal coordinate.

5. General Conics

The term *general conic* should not be confused with the term *general ellipse*, which we just discussed in Section G-4. The use of the name general ellipse was based on the similarity in form to the conventional ellipse with a certain generalization of the exponent. The meaning of general conic was explained at the beginning of this section on conics and encompasses circles, ellipses, parabolas, and hyperbolas, as well as degenerate forms of these functions. We have given special attention to circles and ellipses because of their broad application in design and analysis. For circles we need only three points and/or slope constraints. For ellipses that will satisfy most applications, we have the three or the four constraint version. Normally, if we want a smooth blend or transition, the circular or elliptic arc will be quite adequate among conic options. It is usually not necessary to specify a hyperbolic fit unless the hyperbolic function is the natural underlying behavior pattern of the data. If, however, we should need a five-constraint specification to satisfy a design requirement, then we would solve for the coefficients of a general conic (Eq. (37)), and the solution might, by chance, produce a hyperbola or it might produce one of the other conics. A five-constraint design situation arises for problems as described in Section B and Figs. 5 and 6. There, the objective was to create a curve without inflections that would be tangent to the depicted lines in the figures at P_1 and P_2. Section B described the limitation of fitting a cubic to the four-point/slope constraints. In Section G-3, we say that we can fit a standard ellipse to the constraints. However, we must insure that the slope of the second line is at least twice the slope (in magnitude) of the line joining P_1 to P_2. If this criterion is not met and if the design situation permits the relocation of P_1 and/or P_2 and/or the change in slopes at those points, we may be able to set up an elliptic solution. This process might be cumbersome in certain situations, or we may not have the needed freedom of change. The constraints may be firm or almost so. In such cases, we may be able to supply an input point within the triangle formed by the two lines of Fig. 5 and the line between P_1 and P_2 (Fig. 37). This point is sometimes called a *shoulder* point, and often a designer prefers to add such a constraint. The three points and two slopes now give all the data necessary to solve the general conic equation via five simultaneous equations in five unknown coefficients. The resulting solution may not be a *standard* ellipse or a *standard* hyperbola. In most cases the arc from P_1 to P_2 will be an arc of some ellipse or hyperbola with rotated axes. It can be shown that the five input conditions cannot lie on two branches of a hyperbola as long as the shoulder point is within the triangle. This is particularly fortuitous for a graphic console user because the intermediate

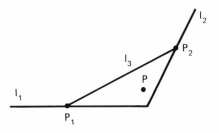

Figure 37 Same as Fig. 5 with added data.

point may be specified by visual pointing. If the resulting curve is displeasing, the operator may move the shoulder point and then view the effect of the change. As in other cases, the operator need not necessarily know the underlying math that produces the curve. His only requirement should be to know whether or not all constraints are satisfied and that the shape meets the more subjective designer preference.

6. Combination of Conics

In Section G-2 we discussed the development of circles, and in Section G-3 we extended the concepts of ellipse development to include the matching or blending of adjoining ellipses. In this section we will describe:

1. The matching or blending of two circular arcs whose radii are, in general, different.
2. The matching or blending of two arcs—one circular and the other elliptic.

These options would be provided to designers where the nature of the application indicates the desirability of using them exclusively or as additional alternatives to those that are considered as the fundamental repertoire.

a. Blending of two circular arcs

As was described earlier in this section, it is often desirable to wrap a continuous curve around a construction, and a single math model may be inappropriate to satisfy design constraints. Two elliptic arcs joined together is one possible approach as was explained in Section G-3. We point out here that the required matching (or tangency) of two circular arcs at a point is more restrictive than the two elliptic-arc case. This may be realized by noting that a range of ellipses from P_1 to P_2, as in Fig. 35, and a corresponding range of input slopes at P_2 are possible within certain constraints. This, of course, creates an option to choose among a range of ellipses. However, only one circular arc with its center on the Y axis can be drawn from P_1 to P_2. Similarly, only one circular arc with its center on the X axis can be drawn from P_2 to P_3. (Centers on the Y and X axes correspond to the same requirement for ellipses—in other words, we have prescribed a horizontal slope at P_1 and a vertical slope at P_3 regardless of the conic curves to be used.)

Because of this restriction, we cannot *require* two circular arcs to be tangent at a *particular* value of P_2. (We used (5, 5) for P_2 in the examples of Section G-3.) However, we may specify one of the coordinates and ask the computer to search for the other. Thus, for the P_1 and P_2 of the examples, we may set the Y coordinate at five units and seek the X coordinates that will produce the desired tangency at P_2. First we will present the technique and the formulas to solve for the X_2 that meets our conditions. We recall in formula (83) that the threshold X_2, $X_2(B)$ for two elliptic arcs is

$$X_2(B) = \frac{4X_3(Y_2 - Y_1)}{3Y_2 - 4Y_1}$$

Since circles are in the family of ellipses, the same formula applies here; so if any solution exists, it must be between $X_2(B)$ and X_3. The procedure to derive the specific X_2 within that range will be described with the aid of Fig. 38.

Using a trial value of X_2 (systematic trials will be described later), we can use the coordinates of P_1 and P_2 to derive the equation of the perpendicular bisector of the P_1P_2 line. This is done by computing the midpoint of that line, the negative reciprocal of the P_1P_2 slope, and a standard line derivation formula. The equation will cross the Y axis at a point we shall call k. Similarly, we can find the X intercept which we shall call h. Using the Y

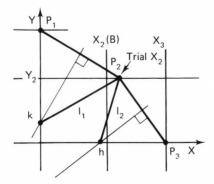

Figure 38 Input data used in searching for solution for two blended circular arcs.

intercept k, and the coordinates of P_2, we determine the slope m_1, of the line, l_1, which connects k to P_2. Since k is at the center of the first arc (because the slope is horizontal at P_1), l_1 is perpendicular to that arc at P_2. Similarly, we can derive the slope of l_2, which we define as m_2. Now for our trial value of X_2 to be a solution, m_1 must equal m_2. We can see pictorially that m_1 is less than m_2 in Fig. 38. Our job is to find the value of X_2 within the prescribed range of possible solutions such that $m_1 = m_2$.

The formulas for m_1 and m_2 in terms of the input coordinate data are

$$m_1 = \frac{Y_2}{X_2} + \frac{X_2^2 - Y_1^2 + Y_2^2}{2X_2(Y_1 - Y_2)}$$

$$m_2 = \frac{2Y_2(X_2 - X_3)}{(X_2 - X_3)^2 - Y_2^2} \tag{85}$$

If we were to set $m_1 = m_2$, we would have to solve for X_2 by means of solving a fourth order polynomial. It is preferable to solve for X_2 by trial and error. For successive values of X_2 from $X_2(B)$ to X_3, $m_1 - m_2$ must start off at $X_2(B)$ with a negative sign (explained in a moment). If the sign does not change with successive trials, no solution exists. If a sign change does occur between two trial values, then a solution exists between the two. We must not lose sight of the fact that the designer does *not* have to concern himself with the actual process. The process, using equations like that of Eq. (85), must be programmed. The designer only has to specify the coordinates of P_1 and P_3 and the ordinate of P_2 and to request the computer to produce two circular arcs which are tangent at P_2. The prescribed trial and error process may be defined in various ways, such as the following:

1. Determine $X_2(B)$. Potential solutions exist if and only if $X_2(B) < X_3$.
2. Divide $X_3 - X_2(B)$ into equal intervals, say ten.
3. At each end point, compute $m_1 - m_2$ to determine its sign.
4. When the sign changes, that interval is subdivided into (e.g. ten) subintervals.
5. The trial procedure is repeated until a change in sign for $m_1 - m_2$ is noted.

6. Then, linear interpolation is applied to derive the desired X_2. This should lead, conservatively, to an accuracy of $1/1000$ of the total trial interval, $X_3 - X_2(B)$.

For data in the first quadrant of a coordinate system, unless $m_1 - m_2$ starts out negative when using $X_2(B)$ as the first trial, there can be no solution because m_1 increases with increasing X_2, and m_2 decreases with increasing X_2. Hence, $m_1 - m_2$ is always increasing. Thus, $m_1 - m_2$ must start as negative if it is going to get to zero (which is needed for m_1 to equal m_2) and produce a solution. Another necessary condition to procure a solution is that m_1 be positive when the trial X_2 is equal to X_3. This would ensure that $m_1 - m_2$ is positive because m_2 is zero when X_2 is set equal to X_3. Thus, a solution can exist if and only if $m_1 - m_2$ is negative when X_2 is set equal to $X_2(B)$, and m_1 is positive when X_2 is set equal to X_3. Using the same data as in Section G-3d on ellipses, which were depicted in Fig. 34, we will go through the same four cases as we did in that discussion. The first case assumes $P_1 = (0, 6)$, $P_3 = (6, 0)$, and $P_2 = (X_2, 5)$. In the elliptic case, we specified P_2 to be $(5, 5)$, but X_2 must be derived for the circular arcs. We will select only a few trial values of X_2 to demonstrate the process.

Using Eq. (83), we determine the boundary X_2 to be

$$X_2(B) = \frac{8}{3}$$

Using $8/3$ as the first trial and employing Eq. (85),

$$m_1 = 1.15$$
$$m_2 = 2.40$$
$$m_1 - m_2 = -1.25$$

Now we use 3 as the next trial. (We will not use many increments of trial X_2 values here because this is merely an illustration of the principle.) For this trial

$$m_1 = 1.33$$
$$m_2 = 1.88$$
$$m_1 - m_2 = -0.55$$

The first two trials indicate that we are proceding toward a solution. By inspection of the trial values and results, we select the next trial to be $10/3$. This gives

$$m_1 = 1.52$$
$$m_2 = 1.50$$
$$m_1 - m_2 = +0.02$$

Thus, $10/3$ for X_2 is very close to the desired solution. Interpolation would yield $X_2 = 3.330$. At that value of X_2, which would be found normally by computer search, the computer can derive two circular arcs from P_1 to $(3.33, 5)$ and from $(3.33, 5)$ to P_3, with tangency at the common point.

As the next case, if $P_1 = (0, 7)$ and $P_3 = (7, 0)$, then a similar procedure produces a solution at approximately $X_2 = 4.8$. As a third case, if $P_1 = (0, 8)$ and $P_3 = (8, 0)$, then the solution is approximately 6.3. This is beyond the $X_2 = 5$, which was used to establish elliptic arcs in Section G-3d. Since we have shown a solution for two circular arcs, we *must* be able to derive a range of solutions for two elliptic arcs (of which the two circular arcs is one of the solutions) if X_2 were set at 6.3 and we followed procedures of Section G-3d.

The purpose of this discussion is to give potential users of such analyses not only the formulas to use but also the "feel" for the parameters that might be changed by manual means at a console and the functions that can be automated. The user of such a system at a console should have no difficulty in utilizing the programmed techniques.

b.   Blending of one circular arc and one elliptic arc

As in the preceding case, the console operator would specify P_1, P_3, and the Y_2 coordinate of P_2. The computer would employ Eq. (83) again to derive the boundary, $X_2(B)$. The designer now selects his preferred X_2 within the scope depicted range, $X_2(B)$ to X_3. This is enough information to generate a single circular arc. If the designer wishes to have a circular arc from P_1 to P_2 (the input data is shown in Fig. 38) to blend with an elliptic arc from P_2 to P_3, the computer will produce the circular arc to X_2 and derive the range of slope options in which a slope must be selected to constitute the elliptic arc. This range will be zero to $1/2 \ (Y_2 - Y_3)/(X_2 - X_3)$, since the elliptic section would be from P_2 to P_3. (The alternative of requesting the elliptic portion from P_1 to P_2 and circular portion from P_2 to P_3 would set the range of slopes at $2(Y_1 - Y_2)/(X_1 - X_2)$ to vertical.) If the slope of the circle at P_2 falls within the acceptable band, the computer will use the slope of the circle at P_2 and derive the blended elliptic arc as well as the circular arc, that is, the total curve from P_1 through P_2 to P_3. If, on the other hand, no solution exists, the designer would be so instructed on the display. This would imply that he would be obliged to increase his input location of X_2 until a solution is achieved or until X_2 reaches the X_3 upper limit. Increasing X_2 for a fixed P_1 and the center on the Y axis, which is our basis, has the effect of increasing the radius of the circular arc and decreasing the slope at P_2 (for the circle/ellipse case). This would move the slope for the elliptic arc towards the acceptable range. Similarly, for the ellipse/circle alternative, we would have to decrease the input X_2 toward $X_2(B)$ in order to make the slope increase towards its acceptable range. Whether or not the designer has the freedom to change the input X_2 is, of course, a function of the nature of the design and the attendant freedom relegated to the designer. Figure 39 depicts an input which will not fulfill the requirements and one that will.

The process of creating a design using conics in a variety of options is a function of many things, including design constraints and designer preferences. The reasons why one may or may not need certain options cannot be completely anticipated. The problem in deciding what options should be a part of the design package is more of an assessment of software development costs and hardware capabilities. It should never be a problem for the designer to use the system. As in this section, the designer decides on some option, such as blending two elliptic arcs around a depicted construction, and so informs the computer. Accordingly, he will provide input (with or without prompting on the display) that is easy to select visually or by keyboard. The computer then shows the options for additional input—for example, the location of $X_2(B)$ for the ellipse/ellipse case. The computer, for example, then computes the range of acceptable slopes and depicts them, and the designer then makes the choice. At this point enough information has been logically gathered, in a true man-machine dialogue, to generate the desired shapes—both for display and for computer retention. If one understands the extremely simple designer procedures (that should be well documented), there is no way (other than software or

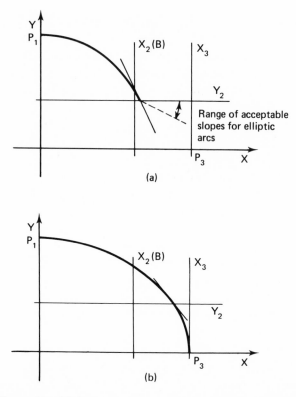

Figure 39 (a) Input X_2 is not large enough to produce circle/ellipse blending at P_2; (b) Input X_2 is large enough to ensure circle/ellipse blending.

hardware problems) that the user cannot proceed to a logical conclusion or solution. Other combinations of math functions are possible and might be developed, as required, along the lines presented in this section.

EXERCISES

1. Can two circular arcs be blended if they have different radii?

2. Is it possible to have data that would permit the blending of the arcs of two standard ellipses but would not have a solution for the blending of two circular arcs? If so, construct a case.

3. Is it possible to have P_1, P_3, and Y_2 such that there is no solution for the blending of any combination of an ellipse and a circle? If so, construct a case.

4. If P_1, P_3, and Y_2 are such that two ellipses cannot be blended, is it possible to find a solution to blend either two circles or a circle with an ellipse?

5. What initial tests can be set up to ascertain whether or not two circles can be blended?

6. If $P_1 = (0, 6.5)$, $P_3 = (7.5, 0)$, and $Y_2 = 5$, derive the smallest X_2 that will permit the blending of two ellipses and/or circles. What is the largest X_2?

7. Using the results of Exercise 6, select an input X_2. Will that selection permit a blend of an ellipse for the upper arc with a circle for the lower arc?

8. What is the relationship between the centers of two blended circular arcs ($m_1 = m_2$) with respect to P_2? Explain why the described process of blending two elliptic arcs does not, in general, apply to the blending of two circular arcs? As a corollary, why is a different process requred for the circle/circle case than the ellipse/ellipse case?

H. EXAMPLES OF OTHER BASIC FUNCTIONS

There are, of course, many other mathematical models and techniques that charaterize certain types of applications. Data derived from a vibrating medium may be well suited to a Fourier analysis. In some cases, one may prefer to use a more general math model, such as rational polynomials, with the objective that such functions contain many of the more basic functions as subsets. However, such generalized systems may have the disadvantage of requiring *too* much input data, as is the case for using a generalized quadratic equation as the base for deriving the equation of a circle.

It should also be noted that while certain math functions might be preferable for certain applications, other basic math functions, such as those for splining, might be moderately well suited with little degradation of efficiency. In this sense, the functions described in the previous sections of this chapter should suffice for a great many applications, even where other functions might be *mathematically* preferable. Notwithstanding these principles, we will present a few other of the more basic but useful functions that are often used for certain types of applications.

1. Exponential Curves

There is no distinction between a logarithmic curve and an exponential curve because

$$Y = e^X$$

is precisely the same function as

$$X = \log_e Y \quad \text{or} \quad 2.303 \log_{10} Y.$$

Therefore, we will arbitrarily discuss the exponential form. The growth or decay phenomena are often exponential in nature. In the mechanical world, it may be asserted, for example, that wear-out or failure rates occur uniformly, that is, the probability of failure is the same for a given time increment. If the probability of failure is 0.1 within Δt, for example, then 10% will fail, on the average, in any Δt span regardless of when it occurs on the total time scale. Another exponential function is the *Gaussian* or *normal* statistical distribution, which is typical of noise distributions, measurement errors, the binomial distribution as the number of trials increase, the distribution of the mean of any distribution, machining tolerances, ballistic range errors, and so forth.

Mechanical failure, for example, may be written as

$$X = \frac{1}{m} e^{-t/m} \tag{86}$$

where Eq. (86) is the relative probability of failure at time t—a probability density function. That is, the area under the curve $0 \leqslant t < \infty$ is unity. The parameter m is the MTBF, *Mean Time Between Failures*. If we integrate zero to one for t, we have

$$P = \int_0^1 \frac{1}{m} e^{-t/m} = 1 - e^{-1/m}$$

which is the probability of failure in the first unit of time. The probability of survival to the end of that unit is, therefore, $1 - (1 - e^{-1/m}) = e^{-1/m}$. In the second unit interval, the probability of failure is

$$P = \int_1^2 \frac{1}{m} e^{-t/m} = e^{-1/m} - e^{-2/m}$$

Forming a ratio of this probability to the probability or percentage survival at the beginning of the second interval, we have

$$\frac{e^{-1/m} - e^{-2/m}}{e^{-1/m}} = 1 - e^{-1/m}$$

which is the probability of failure in the seond time interval.

This demonstrates that the rate of failure is unaltered even though the number of failure decrease—similar to human life expectancy. What does all of this have to do with computing and graphics? Well, suppose we had a lot of failure data and we classified it into intervals of time where failure occurred. The number of failures could be plotted at the center of each classifciation time. By observing the data on the scope, we may wish to see if uniform failure is reasonable (not the case for "infant" failure situations or other situations where wear-out is most likely at some expected time—like light bulbs). If reasonable, we may input trial values of m in Eq. (86) and compute the sum of the squares of the differences between computed and observed values. Trial may be preferable to automatic least squares because of the nonlinearity of the function. That is, the best fit, least squares wise, to an exponential function is not the same as we would get if we were to take the logarithm of the function, make a substitution of variables to linearize the function, and then perform a least squares fit. Thus, trial and error procedures at the scope could be most effective in establishing MTBF and the describing equation.

Depending on the judgment of the analyst or on the nature of the problem, it may be inappropriate to use this exponential form, which implies uniform failure rate. Other math models may be selected as explained in previous sections. In other cases, it may be preferable to use the raw failure data (or some modification of it) for further calculations, such as simulations that require failure rates. For illustrative purposes, we will assume that the following data should be fit with formula (86), and it is necessary to determine the m that gives the best fit.

As an example, suppose the number of classified failures at each of ten successive units of time are observed to be

$$11, 8, 8, 3, 7, 4, 0, 4, 4, 1$$

The computed mean (average) from this data is

$$m = 4.0$$

By dividing each number by 50 (the total number of observations), we reduce each to a fraction. The sum of the fractions will be unity. These fractions can then be compared to those calculated from

$$\frac{1}{4} e^{-t/4}$$

for $t = 1, 2, \ldots, 10$.

The observed and computed fractions are shown in Fig. 40 where the circles represent observed data and the x's represent data computed for $(1/4)e^{-t/4}$ (where 4 is the observed MTBF). The computed values generally fall below those of the observed and will not sum to unity primarily because no point at $t = 0$ is plotted, and the value of $(1/4)e^{-t/4}$ at $t = 0$ is 0.25, where there is no observed classification for this example. Furthermore, the function $(1/m)e^{-t/m}$ would have to be integrated from zero to infinity; so that we are aware that the function would never completely match discrete observed data unless we simply add an offset, which is easy with graphic interaction.

When data is plotted as in Fig. 40, we may visually decide to make some changes and compute some measure of goodness of fit. Such a measure is the standard deviation (each squared difference should be weighted by its classification frequency).

The standard deviation is 0.041 for this example. Let us suppose that the console operator wishes to try alternative values of

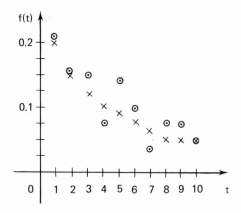

Figure 40   Observed and computed fraction of failures.

m in order to visualize the fit and to compute the standard deviation. If 3.9 were used instead of 4.0 for m, the standard deviation would be reduced slightly to 0.039 from 0.041. Actually, the function is not too sensitive to the value of m (in the region of the mean) in terms of goodness of fit to the data. There are other options that the operator may desire based on visual inspection, such as deleting some data, combining classifications, and local smoothing.

2. Normal Curves

A Gaussian or *normal* curve is of the form

$$F(X) = \frac{1}{\sqrt{2\pi}\,\sigma}\, e^{-1/2[(X-m)/\sigma]^2} \qquad (87)$$

where σ is the standard deviation and m is the mean of the variable X. Equation (87) is a probability distribution function because the integral of $F(X)$ from negative infinity to positive infinity is unity. However, a true normal curve will contain over 95% of the distribution in the range of $X, m \pm 2\sigma$ and over 99% in the range of $m \pm 3\sigma$. The *normal* curve (which is bell shaped about m and is therefore sometimes called the *bell* curve) has many valuable uses in statistical analysis, not the least of which is estimating political vote potentials. However, its use in studying phenomena in a multitude of sciences can be and is being enhanced through its graphical representation and the study of the goodness of fit of data to *normal* curves. The basic behavior patterns of much test data may not be known and may be a candidate for *normal*-curve representation. Conventionally, one computes $\overline{X}$ and S (data average and standard deviation) and uses them for m and σ in Eq. (87). At a graphic console one could then view the plot of Eq. (87) versus the plot of classified data. A standard deviation of error could be computed and analyzed. The operator could vary m and/or σ and view the fit as well as study the analytical aspects. Statistical tests are available to establish tolerance bands, confidence levels, and so on. The geometric portrayal of some of the analytical data can be very useful in understanding the analytical nature of the data and the fitted functions. For example, we know that an unbiased sampling of random numbers from integers zero through nine would be sampling from an underlying population with $m = 4.5$ and $\sigma = 2.86$. A sample of 100 random digits taken from a Rand Corporation random number table is:

65	48	11	76	74	17	46	85	09	50
80	12	43	56	35	17	72	70	80	15
74	35	09	98	17	77	40	27	72	14
69	91	62	68	03	66	25	22	91	48
09	89	32	05	05	14	22	56	85	14

The average of these 100 integers is 4.37; the standard deviation is 2.87. However, statistical theory tells us that the standard deviation of the mean is the same as the standard deviation of the basic population divided by the square root of the sample size ($\sqrt{100} = 10$ in this example); so the standard deviation of the average is approximately 0.29. The 4.37 and the 0.29, (2.89/10), values are well within the sampling tolerance based on a known population (true) mean of 4.5 and a known standard deviation (derived from statistical theory using a discrete variable) of 0.289. Had we not known the true mean, we could use the sample mean to bound or predict the true mean; that is, $4.37 \pm 2(0.29)$ would encompass the true mean with 95% confidence. Much more can be said to make the discussion conform to pure or exact statistical theory, but this should provide a simplified view of some of the important principles—principles that can be studied and expanded by use of graphics. Variations in sample size, alternative fits of data

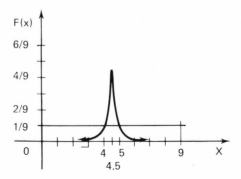

Figure 41   Distribution of random numbers and of sample means.

using alternative m and σ values, inclusion or exclusion of relevant numeric data, and so forth, may be studied. The basic random number probability distribution and the corresponding normal distribution of means of sample size 100 are depicted in Fig. 41.

3. Trigonometric Curves

For certain applications, data is acquired that has cyclical properties such as described by the trigonometric functions. Sometimes they can involve a composite of data representing a variety of frequencies and amplitudes as is represented by a Fourier series. Suppose, for simplicity, we believe that the underlying function is sinusoidal; that is,

$$Y = a \sin bX \tag{88}$$

We wish to determine the parameters a and b that will achieve the best fit to a set of data. If we assume that the best fit implies a least squares fit, then a reasonable approach is to try values of a and b, determine Y in Eq. (88) for each X coordinate of the data, compute the squared difference between each Y from Eq. (88) and the corresponding Y's of the data points, and so forth. This can be greatly assisted by graphics because the console operator may be able to accurately estimate the amplitude, a, and the period, b, by visual inspection. Conversion to the true least squares fit should be rapid through graphic input of trial values. A closed form least squares solution (without iterations) cannot be conveniently developed. However, if we study the discussion of Section B where we showed how well a cubic polynomial can approximate a sine function and if such approximations would suffice, then a closed form (unique solution) cubic least squares approach might be used to fit the data over one cycle. Such alternative procedures for curve fitting may often be employed—the best procedure being dependent on the nature of the application and other factors.

The use of graphics to view the effect of varying combinations of parameters (such as a and b in the present case) can be exceedingly valuable. The analyses that involve many parameters are usually difficult to comprehend if only analytical techniques are used at the exclusion of some form of graphics.

III

THREE-DIMENSIONAL GEOMETRY

There is a considerable requirement to understand the options and techniques of three-dimensional design and analysis. Most designs are of 3-D objects, although some are primarily 2-D with limited 3-D aspects. Generally, there are a number of ways to describe or design an object and, of course, there are many ways to develop analyses that may be characterized by 3-D surfaces. The requirement to develop math models for 3-D objects and surfaces is analogous to that for 2-D. Basically, the development of a math model for an object or a surface is an extension of the concepts and principles of 2-D geometry, but the manifold combinations of techniques and directions that may be taken make the explicit specification of formulas and procedures for 3-D somewhat more nebulous than is the case for 2-D. Nevertheless, a few principles and approaches will be discussed in the following sections.

A. THREE-DIMENSIONAL PRIMITIVES

For applications that require the layout of objects that might be used in architectural configurations, in room or compartment development, and so forth, it may be feasible to use crudely constructed objects that have a resemblance to the "real-life" objects they are supposed to depict. Such objects may be constituted at a graphic terminal by piecing together certain "building-block" primitives of which the geometric definitions and physical and/or descriptive attributes are known. The complexity and number of primitives that might be used efficiently depend on the extent to which an application can tolerate inexactness in object representation. For example, would we need a refined definition of a chair with all of its curved contours or might we find that a composite of rectangular boxes of different dimensions would suffice? Naturally, the fewer and simpler the primitives are, the better. Too large a set would likely contain some elements that are difficult to represent mathematically. It is more difficult to search for the best elements to use when the set is large, and we are not likely to get a precise design no matter how many elements are in the set unless we are dealing with simple designs. We found that six primitives of simple construction would serve one particular application quite adequately in developing ship's compartments. These primitives, each of which can be defined by just a few computer words, were rectangular boxes, right circular cylinders, spheres, wedges, cones, and frusta of cones. They are depicted by three views in Fig. 42a and by three views with rotation in Fig. 42b.

Nine computer words will define the rectangular box—length, width, height, displacement in X, Y, and Z, and three angles of rotation. Each rectangular primitive has the six degrees of freedom for translation and rotation plus the basic descriptors. A cylinder requires a radius and a height, a frustum of a cone requires two radii and a height, the complete cone requires one radius and a height, the sphere requires only a radius, and a wedge requires length, width, and height. A selection of one of the six primitives and input of the dimensions and the six position parameters will place any of the primitives anywhere in space. Much less data is required to describe each primitive than would be the case for pure data representations. Users of such primitives assert that there is a short learning period involved in positioning primitives to desired angles, but after that, console users have no trouble in positioning them wherever they like. Of course, the actual program to construct a primitive has to be developed. It is an easy task to combine primitives to form shape approximations of objects, such as turrets,

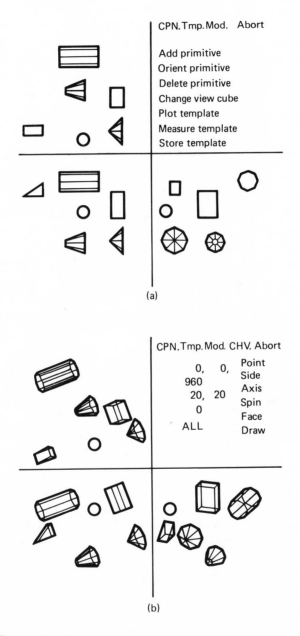

Figure 42 (a) Building block primitives in three views; (b) Primitives with rotation.

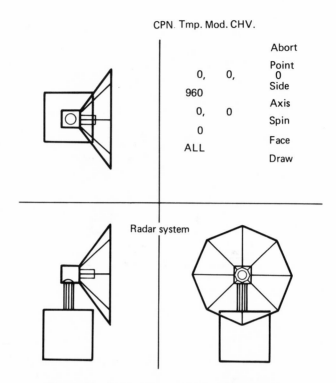

Figure 43   Objects formed from 3-D primitives.

radars, engines, chairs, and tables. Figure 43 depicts examples of primitive construction.

It should be noted that some primitive constructions, such as cylinders, have additional lines drawn to facilitate viewing regardless of the viewing angle. Such constructions lessen the requirement to find boundary contours or hidden lines for each presentation. Mathematical determinations of boundaries and hidden lines are relatively complex and impose somewhat of a burden on programming and CPU facilities. Although such determinations may be essential for some applications, they should be avoided when there are viable alternatives as exemplified by the use of primitives. Figure 44 shows the top and front views of a cylinder that has been constructed in standard position with only two basic vertical lines. Figure 45 shows the same cylinder rotated 90° about its vertical axis. Thus, we see the effect of using insufficient basic lines to represent a 3-D object when using primitives.

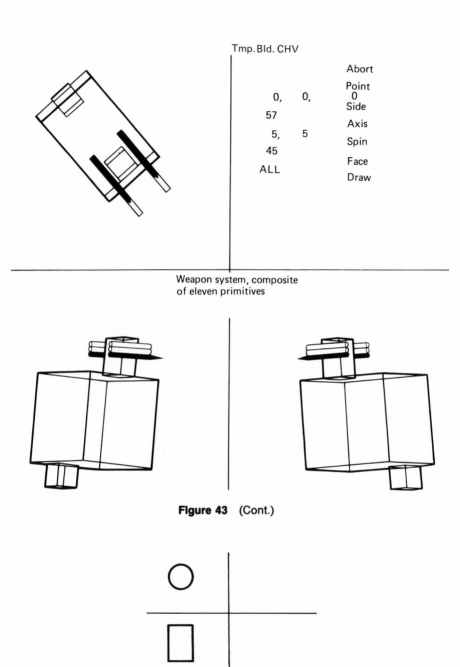

Tmp. Bld. CHV

Abort

Point

0, 0,   0

57   Side

            Axis

5,   5   Spin

45

            Face

ALL

            Draw

Weapon system, composite
of eleven primitives

Figure 43   (Cont.)

Figure 44   Cylinder in standard position.

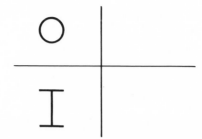

Figure 45 Cylinder rotated 90° showing effect of insufficient construction.

B. DISTINCTION BETWEEN 3-D DRAWINGS AND 3-D PRIMITIVES

At this point, it is reasonable to try to understand the fundamental delineation between the use of 3-D primitives to represent 3-D objects and the use of conventional drafting techniques—the creation of line and curve drawings in orthogonal views. There are software systems that are commercially available that perform many or most of the elements of drafting. There are, of course, many applications that can make use of a graphic drafting package because it is a drawing that feeds the vital pictorial and dimensional data to the fabrication phase. The manufacturer of an article can comprehend a great deal from a multiview drawing. Furthermore, a system that provides comprehensive line- and curve-drawing facilities in any 2-D reference plane provides the basis for depicting a complex and detailed part or object. This cannot be so readily accomplished with primitives because of the limited repertoire that must characterize any set of primitives in the practical sense. However, the primitive has one distinctive property that is missing in the drawing—the full 3-D definition. We can rotate and otherwise move primitives and combinations of primitives at will in 3-D. We cannot, in general, do that with a drawing that is no more than a composite of vectors representing some object. If, for example, we have orthogonal views of an object with dotted lines to represent hidden contours, we can visualize the shape of the object of which the views are drawn; but the computer cannot automatically survey two views and impart 3-D integrity in the sense of being capable of repositioning the depicted object and calculating its physical proper-

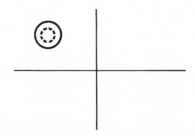

Figure 46 Top view of one or more objects.

ties. Some systems give more potential to drawings by giving the console operator a procedure by which drawing boundary curves can be systematically indicated to represent boundaries of surfaces which the computer then automatically develops. This is done by any of a number of interpolation procedures that must be programmed into the process. This enhanced drawing technique is a more general alternative to the use of a specific set of primitives and can be used as a method of creating primitives per se. The process is, however, considerably more complex. To illustrate the difficulty in automating the full 3-D object definition from two orthogonal views of a drawing, consider Fig. 46, which is the top view of one or more objects.

Can the reader tell what the object is from this one view? Presumably, he cannot, because the one view cannot define a unique object. Now consider Fig. 47, which shows the front view of three different objects, each of which has the same top view as shown in Fig. 46.

There are, of course, many other objects that could have the same top view. The point is, one view does not define an object and two views will not tell the computer what it is. In some cases, even two views will not define an object. If we wished to rotate the objects in Fig. 47, 45° about a horizontal axis in the tops, Fig. 47 would become Fig. 48.

The three cases of Fig. 48 depict what the three objects would actually look like if rotated about an axis in the top. Dotted curves represent hidden contours that would be observed if the objects were transparent. Such drawings could be constituted at a drawing board or at a console that has a programmed drawing capability. However, it is important to understand that had the drawing been made as in Fig. 47, the computer would not be able to perform rotation of the objects and produce the displays depicting Fig. 48.

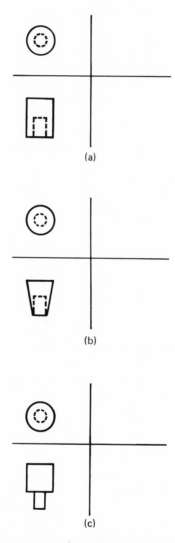

Figure 47 (a) Cylinder with a hole in the bottom; (b) Frustum of a cone with a hole in the bottom; (c) Cylinder with a peg on the bottom.

It could rotate *only* those vectors which constitute Fig. 47, and this would not accommodate changing boundaries and hidden curves that represent the 3-D objects. This is the fundamental distinction between making 2-D drawings of an object and defining an object in 3-D and having the computer derive 2-D projections in any desired direction. Two final points pertaining to drawings and to

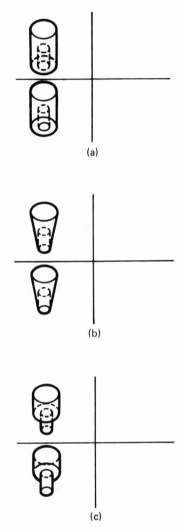

Figure 48 (a) Fig. 47(a) rotated 45° about axis in top (bottom toward the viewer); (b) Fig. 47(b) rotated 45° about axis in top; (c) Fig. 47(c) rotated 45° about axis in top.

the use of primitives come from the fundamental principles just described. First, drawings of objects that are bounded by geometric planes can be more easily translated, rotated, scaled, and so on, just like primitives, as long as all corners are specified in the different orthogonal views. However, even then the defining procedure has to be carefully executed.

The second point is that each of the three objects of Figs. 47 and 48 can be constructed by using two of the primitives described in Section A. To extract the most utility from the primitives, construction should include the facility to use a negative primitive, that is, to subtract a primitive. In the first two cases of Figs. 47 and 48, the cylindrical hole can be subtracted from the larger cylinder. However, the portrayal of hidden lines is by no means automatic. What can be made automatic with primitives is the portrayal of construction lines and curves as if the object were only a wire-like frame, which is equivalent to having a transparent object.

The principles set forth in the first two sections of this chapter should be carefully studied in order to make a meaningful assessment of the virtues and limitations of commercially available 3-D graphic packages and to make decisions regarding in-house software development. It is exceedingly important to comprehend the distinction between the representation of an object or a surface by 2-D drawings or by logical build-up of the 3-D definition via algorithms or via use of primitives. Algorithms, built into the system, permit automatic 3-D development while the graphic user works in a normal 2-D mode. This results in an enriched data base to be used for many important subsequent processes such as: the derivation of additional drawings; ease of major design changes or refinements; computation of areas, volumes, surface intersections, etc.; development of analyses related to the geometry; and the integration of all computerized processes.

C. 3-D BOUNDARY CURVES FOR SURFACES

The necessity to develop curves in 3-D is often predicated on the need to generate surfaces. Such curves may be used as boundary curves, and various interpolation schemes or algorithms can be prescribed that will define a surface of some kind. The surface may be a physical shape that represents some portion of an object of interest, such as an automobile panel or a container, or it may be a portrayal of parameters in 3-D, such as pressures on a plate, volume versus temperature, and pressure of a gas. Thus, a surface may be used in design, analysis, or in the visualization of different types of data. In an analytical mode, data might be derived from instrumentation. Such empirically derived data may be quite erratic because of measuring difficulties or the influence of extraneous unwanted parameters that cannot be filtered out. The choice of curves to fit the data, the judicious use of smoothing procedures, the curve-fitting technique, and the methods used to extend the data to represent surfaces are important factors in developing surfaces that will impart accurate meaning when they are used in analysis.

1. Space Curves

In a design mode, the development of 3-D curves may be predicated on design constraints or criteria. In some cases, the boundary curve may be a space curve defined either by a set of 3-D coordinates or by design constraints. In either case, the space curve and its projections might be portrayed as in Fig. 49.

Figure 49 shows a space curve from point A to point E with 3-D projections from A to C, B to E, and D to E. If the space curve AE is constituted from a set of data points, the analyst has several options. As was discussed in Section E (Parametric Techniques) of Chap. II, there is an option to fit each variable X, Y, and Z as independent functions of arc length, where arc length is approximated by the successive chords from data point to data point. Alternatively, one of the three variables, say X, might be used in the same way that a parameter would be used. Thus, two curves would result (AC and BE) representing Z and Y as functions of X; that is, as X is varied through its range of values, Z and Y will be

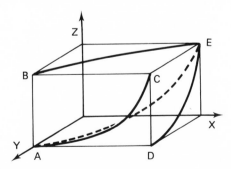

Figure 49 Space curve with projections in three orthogonal planes.

generated. The different ways to fit the 2-D relationships are explained in Chap. II. They include the use of least squares smoothing (with some basic underlying function such as a cubic polynomial) and the use of splining. Such 3-D data may arise from
many sources as exemplified by the digitizing of shapes, such as
automobile models and the readings from contour maps. Whatever
the source, the implication is that the fitted data will be used for
subsequent design and/or analysis tasks.

Regardless of the fitting option, the representation of 3-D data
by multiple 2-D math models will not fit the 3-D shape precisely.
However, for smooth or gradually changing data, the approaches
just suggested will usually be quite adequate and are considered to
be practical. Should that not be the case, additional data points (if
obtainable) should improve the representation. It is not at all un-
common to have a 3-D curve that is not defined or represented by a
set of data points but does have logical design constraints. For
example, a width boundary (called the *half-breadth* line) of an
airplane fuselage may rise vertically and move closer to the airplane
center line as it recedes towards the tail. Figure 50 shows a top view
and a side view of the half-breadth (H-B) contour of the rear section
of a sample fuselage.

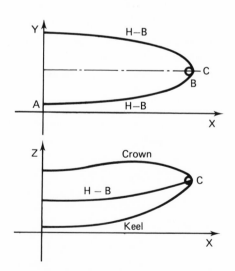

Figure 50 Top and side view of an airplane fuselage
section.

The 3-D half-breadth space curve can be adequately defined in early design stages by deriving $Y = f(X)$ and $Z = g(X)$, two functions of X, from the value of X at the constant cross section to the value of X at the aft extremity. Since airplane fuselages may not shrink to a needle-like point at the end, the designer may specify circles (not necessarily of the same radii) to which the aft contours will become tangent. The requirements for Y and Z as functions of X might be:

1. For Y, have tangency at point A and at point B, then follow the circle to point C.
2. For Z, have tangency at point A and terminate at point C.

In this simple example, the tangencies or slopes at A will be zero. In the development of $Y = f(X)$, there are four input conditions—the coordinates of A and B (in X and Y) and the slopes at each point. We might prescribe an elliptic arc (from an ellipse in standard position) fit from A to B. To ascertain the slope and coordinates at B, the console operator would specify, perhaps with a light pen or other input device, the position of point B. The slope would be that of the circular arc at B. Then the computer can derive an elliptic arc for $Y = f(X)$ as long as the slope at B meets the necessary requirements as described in Section G (Conics) of Chap II. If the necessary condition is not met, the designer may move point B closer to C until a solution is found and the arc is plotted on the display. The designer will pass personal judgment on the suitability of the resulting curve that is developed through the man-computer dialogue. Similarly, a curve for $Z = g(X)$ can be developed to match the two points and one slope that is innate to the design. Those three conditions would be mathematically adequate to derive a parabolic or circular arc, although the nature of the application would preclude the circular arc. If some other curve type for Z is preferred by the designer, say a standard position ellipse, then another constraint must be specified, such as an intermediate point or a slope at the terminus. If more control is desired on the curves for either Y or Z, then a fifth condition may be specified, and this could produce a general conic if that math form is requested. It should be noted that there could be a number of curve types to choose from as discussed in Chap. II, but the nature of the design would preclude cubic polynomials as feasible options because inflection is undesirable for the H-B contour.

It should be observed that this type of design example is not conducive to acquiring numerous 3-D points to represent the H-B contour. The basic shape dictates certain point and slope constraints with some freedom on the part of the designer to add his own constraints to achieve a boundary curve that meets both functional and aesthetic requirements. A spline through approximated points would produce unwanted inflections.

2. Curves in a General Plane

a. Drawing in a 2-D projection plane

From a design or styling point of view, it may be desirable to determine what kind of 3-D curve in a plane is needed to produce a desired curve or shape in one or more planes of projection. Suppose, for example, an object is to be designed that has certain faces (planes) and 3-D curves that bound them. It is desirable to have a particular boundary that projects, according to some prescribed path, in the XY plane. This is depicted in Fig. 51.

To accomplish the objective, it is merely necessary to establish a curve in a 2-D plane as AB is established in XY in Fig. 51. The method of establishing a 2-D curve is the subject of the previous chapter. With such a curve and the equation of the plane that is to contain one of the object's 3-D boundary curves, it is merely a matter of solving for Z in the plane for each pair of XY coordinates that are selected from the 2-D equation or curve (exemplified by AB in Fig. 51). The only new requirement for setting up the solution is the determination of the equation of the plane to which the points in XY (or another orthogonal plane) will be projected. Since many analytical texts provide the method or procedure to get the equation

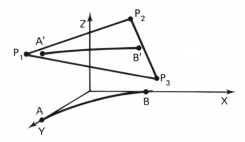

Figure 51   Boundary curve, A′B′, as a function of projected curve, *AB*.

of a plane but do not provide the resulting formulation, one formulation will be given here. Suppose the object to be designed is to have a boundary in a plane defined by three points: $P_1(X_1, Y_1, Z_1)$, $P_2(X_2, Y_2, Z_2)$, and $P_3(X_3, Y_3, Z_3)$ as in Fig. 51. The general form of the equation of a plane is

$$AX + BY + CZ = D$$

We will substitute A, B, C, and D in terms of K coefficients as follows:

$$K_{10}X + K_9Y + \frac{K_8}{K_7} Z = 1 \qquad (89)$$

The coefficients of Eq. (89) are derived by the following set of formulas:

$$
\begin{aligned}
K_1 &= X_2Y_1 - X_1Y_2 \\
K_2 &= X_2Z_1 - X_1Z_2 \\
K_3 &= X_2 - X_1 \\
K_4 &= X_3Y_1 - X_1Y_3 \\
K_5 &= X_3Z_1 - X_1Z_3 \\
K_6 &= X_3 - X_1 \\
K_7 &= K_4K_2 - K_1K_5 \\
K_8 &= K_4K_3 - K_1K_6 \\
K_9 &= \frac{K_3 - K_2K_8/K_7}{K_1} \\
K_{10} &= \frac{1 - K_9Y_1 - (K_8/K_7)Z_1}{X_1}
\end{aligned}
\qquad (90)
$$

There may be a default condition such as would occur if X_1, K_1, or K_7 were zero. Such cases must be provided for, which is more of an exercise in detailing than in principle; so we will skip it for this discussion. Now, we will take an illustration of the principles just presented. Suppose, for whatever reason, we wish to develop a curve in a plane that will have a projection in the XY plane of $Y = 4 \sin (\pi/4)X$. Three points on the oblique plane are P_1 (1, 2, 4), P_2 (3, −1, 5), and P_3 (4, 0, 6). Using these three points and Eq. (90) leads to the planar equation,

$$4X + Y - 5Z - 14 \qquad (91)$$

This equation can be derived rapidly by manual means and even more so with a pocket calculator. Therefore, it is recommended that the reader verify Eq. (91) by applying Eq. (90) to the three points. If we evaluate $Y = 4 \sin (\pi/4)X$ for selective X's from zero to four and substitute in Eq. (91), we will derive the 3-D curve in the oblique plane that produces the sinusoidal projection in the XY plane. The tabular coordinates are as follows:

X	$Y = 4 \sin(\pi/4)X$	Z (from Eq. (91))
0	0	2.80
1	2.83	4.17
2	4.00	5.20
3	2.83	5.77
4	0	6.00

The plots of the desired 3-D curve and its sinusoidal projection are shown in Fig. 52. From the data in the table (and additional tabulated points if needed), projections in other views can easily be obtained, that is, YZ and XZ. In the XZ plane, for example, the tabular points lead to Fig. 53.

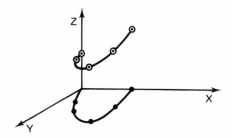

Figure 52 Creating a 3-D curve in a plane that produces a specified projection.

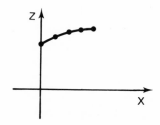

Figure 53 Curve of Fig. 52 projected into the XZ plane.

A mathematical equation for the plot of Fig. 53 could be obtained by substituting $Y = 4 \sin (\pi/4)X$ into the plane represented by Eq. (91). However, in many cases the XY (or other) projection equation is unknown or too cumbersome to work with via substitution. In such cases, actual coordinate data may be developed to get a 2-D relationship and project it to the plane as in the table. The standard 2-D curve fitting may be employed when and if required. Thus, we may use an empirical relationship of Y versus X. We may fit a curve or use interpolation and/or extrapolation to get additional (XY) values. The (XY) values will be substituted in the derived plane to acquire corresponding Z coordinates. A curve fit of Z versus X, Y, or arc length may be employed to acquire a math model for Z if it is deemed desirable.

EXERCISES

1. It is desired to work in a plane defined by $(-1, -1, -4)$, $(2, -2, 3)$, and $(4, 0, -4)$. What is its equation?

2. It is also desired that a curve in the plane of Exercise 1 have a projection that is circular in the YZ projection with equation $Y^2 + Z^2 = 16$. Find the trace of XYZ on the plane for YZ coordinates in the first quadrant of the YZ plane. Plot X versus Y. What kind of shape results? Why?

3. Suppose the data in the XY projection plane has coordinates $(0, 0)$, $(2, 1)$, $(4, 4)$, $(6, 5)$, $(8, 5)$, $(10, 3)$, and $(12, 0)$. Using the plane of Exercise 1, find the corresponding Z coordinates. Plot the given Y versus X and the derived Z versus X.

b. Drawing in a general plane

For many applications, it is required to develop a function or a boundary curve in an oblique plane—a plane not parallel to the principal coordinate planes. This differs from the preceding discussion in the sense that what is desired is a specfic shape in 3-D, not a specific shape in a projection plane. If, for example, we would like to draw an elliptic arc of certain constraints in the plane of Eq. (91), we could not do so by first developing a curve in the 2-D XY plane and then by projecting those XY data onto the plane of Eq. (91). Therefore, we must follow the reverse process; that is, draw the desired curve directly in the plane and then project it to any desired view in 2-D, such as that in the XY plane. To do this, consider a plane represented by three points as in Fig. 54.

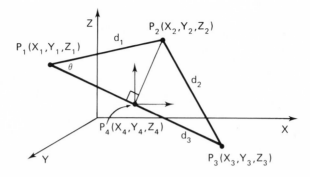

Figure 54 Plane in 3-D in which boundary curves will be constructed.

It is desirable to establish an origin in the $(P_1P_2P_3)$ plane, and this will be designated $P_4(X_4, Y_4, Z_4)$. This is done via the following procedure:

1. Using direction cosines and the formula for an angle between two lines, we have

$$\cos \theta = \frac{(X_2 - X_1)(X_3 - X_1) + (Y_2 - Y_1)(Y_3 - Y_1) + (Z_2 - Z_1)(Z_3 - Z_1)}{d_1 d_3}$$

where d_1 and d_3 are the distances between P_1 and P_2 and P_1 and P_3, respectively.

2. The distance from P_1 to P_4 would then be

$$d_1 \cos \theta$$

3. Using proportions, the P_4 coordinates would be

$$X_4 = X_1 + \frac{d_1 \cos \theta}{d_3} (X_3 - X_1)$$

$$Y_4 = Y_1 + \frac{d_1 \cos \theta}{d_3} (Y_3 - Y_1) \qquad (92)$$

$$Z_4 = Z_1 + \frac{d_1 \cos \theta}{d_3} (Z_3 - Z_1)$$

Having derived the coordinates of P_4, we may redraw the $P_1P_2P_3P_4$ plane in 2-D, and we define the abscissa and ordinate as U and V, respectively. This is depicted in Fig. 55.

Now we can construct any shape we desire in the UV coordinate system and then, with appropriate transformations, change the

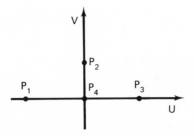

Figure 55 The oblique plane of Fig. 54 redrawn in 2-D.

data points in UV to coordinates in XYZ. This procedure is, of course, well suited to computer graphic principles. Let us assume that we wish to design an object that has one planar face in the $P_1P_2P_3$ plane, which is now the UV system as in Fig. 55. Now each data point in UV must be transformed as previously stated. Suppose W is an axis perpendicular to U and V. For a general 3-D space curve, we would convert UVW coordinates to $U'V'W'$, which are axes parallel to X, Y, and Z, respectively.

Let the angles between U and U', V', and W' be α_1, α_2, and α_3, respectively. Let the angles between V and U', V', and W' be β_1, β_2, and β_3, respectively. And, let the angles between W and U', V', and W' be γ_1, γ_2, and γ_3, respectively.

For these definitions and previous input or derived data, we can get the necessary direction cosines as follows:

$$\cos \alpha_1 = \frac{X_4 - X_1}{d_1 \cos \theta}, \qquad \cos \beta_1 = \frac{X_2 - X_4}{d_1 \sin \theta},$$

$$\cos \alpha_2 = \frac{Y_4 - Y_1}{d_1 \cos \theta}, \qquad \cos \beta_2 = \frac{Y_2 - Y_4}{d_1 \sin \theta},$$

$$\cos \alpha_3 = \frac{Z_4 - Z_1}{d_1 \cos \theta}, \qquad \cos \beta_3 = \frac{Z_2 - Z_4}{d_1 \sin \theta},$$

$$\cos \gamma_1 = \pm\left(1 - \cos^2 \alpha_1 - \cos^2 \beta_1\right)^{1/2}$$

$$\cos \gamma_2 = \pm\left(1 - \cos^2 \alpha_2 - \cos^2 \beta_2\right)^{1/2}$$

$$\cos \gamma_3 = \pm\left(1 - \cos^2 \alpha_3 - \cos^2 \beta_3\right)^{1/2}$$

The proper sign for each $\cos \gamma$ is determined by the following convention. In setting up the UV system, choose the smallest X and the largest X as the location of points P_1 and P_3, respectively. Then there are four cases:

1. When P_2 is above the P_1P_3 line (in Y) and in front of it (in Z), $\cos \gamma_1$ is negative, $\cos \gamma_2$ is negative, and $\cos \gamma_3$ is positive.
2. When P_2 is above the P_1P_3 line and behind it, $\cos \gamma_1$ is negative, $\cos \gamma_2$ is positive, and $\cos \gamma_3$ is positive.
3. When P_2 is below the P_1P_3 line and in front of it, $\cos \gamma_1$, $\cos \gamma_2$, and $\cos \gamma_3$ are all positive.
4. When P_2 is below the P_1P_3 line and behind it, $\cos \gamma_1$ is positive, $\cos \gamma_2$ is positive and $\cos \gamma_3$ is negative.

(These formulas are based on the assumption that P_1 to P_3 is the U axis, that the positive V axis is above P_1P_3, and the positive W is forward of P_1P_3.) Then the desired transformations are

$$U' = U \cos \alpha_1 + V \cos \beta_1 + W \cos \gamma_1$$
$$V' = U \cos \alpha_2 + V \cos \beta_2 + W \cos \gamma_2 \qquad (93)$$
$$W' = U \cos \alpha_3 + V \cos \beta_3 + W \cos \gamma_3$$

which convert any point in a UVW system to a system parallel to XYZ. This is a general formula and represents the transformation for a 3-D curve. However, in the case of this section, all points are generated in U and V with $W = 0$. Therefore, for 3-D curves in a plane, Eq. (93) may be rewritten:

$$U' = U \cos \alpha_1 + V \cos \beta_1$$
$$V' = U \cos \alpha_2 + V \cos \beta_2 \qquad (94)$$
$$W' = U \cos \alpha_3 + V \cos \beta_3$$

Finally, we derive the XYZ coordinates by simple translation, that is,

$$X = U' + X_4$$
$$Y = V' + Y_4 \qquad (95)$$
$$Z = W' + Z_4$$

Now we will carry through an example. We wish to describe the curve of Eq. (97) (presented in a moment) in a particular plane and then show it in an XYZ system. Suppose we have three points that define a plane. We will use the points of the example that led to Eq. (91). (The equation of the plane is not needed in the current case.) Thus, $P_1 = (1, 2, 4)$, $P_2 = (3, -1, 5)$, and $P_3 = (4, 0, 6)$. We refer to Eq. (92) and its subsidiary equations to derive P_4, the origin of UV as expressed in XYZ coordinates. Thus

$$d_1 = \sqrt{14}, \qquad \cos\theta = 0.907$$

$$d_2 = \sqrt{3}, \qquad X_4 = 3.47$$

$$d_3 = \sqrt{17}, \qquad Y_4 = 0.35$$

$$Z_4 = 5.65$$

Then we may represent P_2, P_3, and P_4 in a 2-D UV system as in Fig. 55. Referring to the subsidiary formulas for Eq. (93), we derive

$$\cos\alpha_1 = \quad 0.731, \quad \cos\beta_1 = -0.299$$

$$\cos\alpha_2 = -0.483, \quad \cos\beta_2 = -0.860$$

$$\cos\alpha_3 = \quad 0.483, \quad \cos\beta_3 = -0.414$$

Then from Eq. (94)

$$U' = \quad 0.731\,U - 0.299\,V$$
$$V' = -0.483\,U - 0.860\,V \qquad (96)$$
$$W' = \quad 0.483\,U - 0.414\,V$$

These equations permit us to convert any curve (or coordinates of points on a curve) in the UV plane (which contains P_1, P_2, P_3, and the derived P_4) to points in a rotated system that is parallel to the XYZ system. Then, Eq. (95) completes the transformation. To carry on the example, we will assume that we wish to develop a boundary curve of an object that has a planar face in the UV system. For expediency, we will use only the first quadrant plot of the hypothetical boundary and we will define it by the equation

$$\frac{U^{1/2}}{(P_4 P_3)^{1/2}} + \frac{V^{1/2}}{(P_4 P_2)^{1/2}} = 1 \qquad (97)$$

The distances from P_4 to P_3, $(P_4 P_3)$, and P_4 to P_2, $(P_4 P_2)$, are 0.725 and 1.57, respectively, as determined by using the standard Pythagorean distance formula. Thus, Eq. (97) becomes

$$\frac{U^{1/2}}{0.725^{1/2}} + \frac{V^{1/2}}{1.57^{1/2}} = 1 \qquad (98)$$

and its graph is as depicted in Fig. 56.

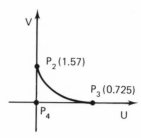

Figure 56 Plot of Eq. (98) in an oblique plane projected in 2-D.

A tabulation of the UV data derived from Eq. (98), the $U'V'W'$ data derived from Eq. (96), and the XYZ equivalent derived from Eq. (95) are:

U	V	U'	V'	W'	X	Y	Z
0	1.57	−0.47	−1.35	−0.65	3.00	−1.00	5.00
0.15	0.47	−0.03	−0.48	−0.12	3.44	−0.13	5.48
0.30	0.20	+0.16	−0.32	+0.06	3.63	+0.03	5.71
0.45	0.07	+0.31	−0.28	+0.19	3.78	+0.07	5.84
0.60	0.01	+0.44	−0.30	+0.29	3.91	+0.05	5.94
0.725	0	+0.53	−0.35	+0.35	4.00	0	6.00

It should be noted that the first and last sets of X, Y, and Z coordinates are the original P_2 and P_3—two of the three points that define the plane in which the boundary curve is drawn. The tabulation shows the UV coordinates of a curve drawn in a plane and the resulting XYZ coordinates. Therefore, after designing a planar boundary curve, one may view the results in any way desired by reference to an XYZ coordinate system of his choice. In our example we can easily plot projections in any of the coordinate planes. The YZ projection of the planar curve of Fig. 56 is shown in Fig. 57.

It is important to understand that the *user* of this procedure need not understand the theory and the transformations that convert input data to output displays. This is the case as long as the formulations are programmed correctly. A display console operator would input three space points (or equivalent defining constraints). The computer would display the plane on the screen in 2-D. The

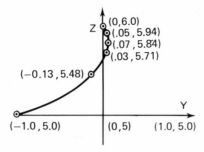

Figure 57 Planar curve of Fig. 56 projected onto the YZ plane.

operator would then create a 2-D curve or 2-D data plot as described in Chap. II. The computer completes the job and will then present a plot in any view requested by the operator.

It may appear that the two techniques discussed (drawing in a plane and projecting) are essentially the same. There are subtle differences, which we discussed in the introductory remarks of the two techniques. However, the technique just discussed is quite general and will permit the transformation of a 2-D drawing in any plane to any set of 3-D orthogonal axes. It is merely necessary to establish a coordinate system in the plane and then establish the angles between the axes of the two systems. Equations (92) through (95) and their subsidiary equations are sufficient to accommodate this objective. More generally, Eq. (95) permits us to transform a general space curve.

EXERCISES

1. Because of the tilt of an airplane engine with respect to the basic coordinate axes, it is desired to construct an elliptic profile in a plane. Three points that define the plane are $P_1(0, 0, 3)$, $P_2(3, -2, 6)$, and $P_3(6, 0, 0)$. It is desired that the elliptic arc be tangent to the line P_1 to P_2 at P_2, that the arc terminate at P_3, and that the reader will supply the input slope of his choice at P_3 after the plane has been transformed to a UV 2-D system. Plot the projections of the elliptic arc in each of the principal XYZ planes.

2. Using the UV system of Exercise 1, draw a circle with radius 4 and center at $(2, 3)$ in UV. Plot the projections of that circle in each of the principal XYZ planes.

D. SURFACES FORMED FROM FOUR SPACE CURVES—
AT LEAST TWO IN A PRINCIPAL COORDINATE PLANE

The mathematical models that describe a surface can be constituted in many ways. Therefore, it is more appropriate to present
techniques and methodology than to attempt to present formulas as
we did in Chap. II. When possible it is certainly desirable to use
known mathematical formulas such as those for cylinders, spheres,
ellipsoids, and cones. More generally, a surface may be regarded as
being developed from an interpolation and/or extrapolation process
as applied to boundary points and curves. This concept gives a
much broader range of options. In many applications it is convenient to describe 2-D boundary curves in planes parallel to the
principal coordinate planes. As an example, suppose we wish to
define the surface of an inlet that has special properties as shown in
Fig. 58. We specify that the surface cross section should start off
circular and then expand while changing shape to a final cross
section that approaches a rectangle.

We note from Fig. 58 that the surface will be defined over a
range of Y from zero to three. At $Y = 3$, both X and Z intercepts
have expanded to the dimension 2. At that cross section, we desire
our curve to pass through a shoulder point that is located at 0.9 of
the length of the diagonal of the square shown in the figure at

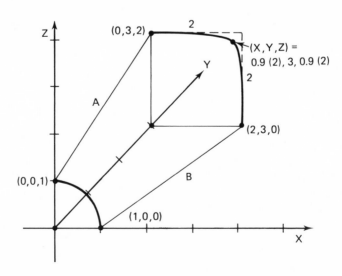

Figure 58 Surface from changing cross section.

$Y = 3$. This kind of definition is consistent with the concept of generalized ellipses as described in Chap. II, Section G-4. Thus, we may describe the surface as a continuous change in cross section from the circle at $Y = 0$ to the generalized ellipse at $Y = 3$. The circle has an equation

$$X^2 + Z^2 = 1$$

This is in the family of generalized ellipses that passes through a shoulder point of the diagonal at $\sqrt{2}/2$ or 0.707 of the diagonal. Using the coordinates of $(0, 2)$ and $(2, 0)$ for P_1 and P_3 (Chap. II, Section G-4) and a 0.9 fraction of the diagonal, we derive the equation

$$X^{6.579} + Z^{6.579} = 2^{6.579} \quad \text{at } Y = 3$$

Our objective is to derive the describing equation that relates Z and X for any value of Y between its limits of zero and three. The question is how do we interpolate to arrive at the cross section equation for any Y. Using generalized ellipses, we must find the exponent n as a function of Y. We must also define the X and Z intercepts as a function of Y, as represented by curves A and B in Fig. 58. The latter are derived by the describing equations:

$$\text{For } A, \qquad X = 0, \qquad Z = \frac{1}{3}Y + 1$$

$$0 \leqslant Y \leqslant 3$$

$$\text{For } B, \qquad Z = 0, \qquad X = \frac{1}{3}Y + 1$$

As for the value of n, there are two interpolation alternatives:

1. Interpolate on n between $n = 2$ and $n = 6.579$.
2. Interpolate on the fraction of the diagonal, between 0.707 and 0.9.

This is a judgment decision that someone has to make. However, the second alternative seems to represent the most direct physical situation; the shoulder point is gradually changing, and we seek the equivalent equation variation—not vice versa. Whichever alternative we select, we still must make a judgment on what interpolation process should be followed to characterize the cross section transition. Selecting the fraction of diagonal variation, we will assume that the cross sections at $Y = 0$ and at $Y = 3$ are constant immediately beyond these limits. Thus, the fraction of the diagonal may be represented as in Fig. 59.

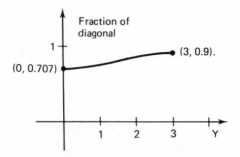

Figure 59 Variation in location of a shoulder point as a fraction of the diagonal.

A cubic polynomial fits this situation very well. We have four conditions that lead to the equation as was explained in Chap. II, Section B on polynomials. We have two points, (0, 0.707) and (3, 0.9), and the design assumptions lead to slopes of zero at both $Y = 0$ and $Y = 3$. (The derivation of the describing equation is simplified considerably if we let the variation in Y be between zero and unity. This is admissible as long as any assigned value of Y is converted to a fraction of the interval of concern.) With this normalization ($0 \leqslant Y^* \leqslant 1$) the cubic equation becomes

$$F = -0.386 Y^{*3} + 0.579 Y^{*2} + 0.707$$

where F is the interpolated fraction of the diagonal. This equation gives F for any fraction of Y^* in the interval of concern. For illustrative purposes, we select 1/4, 1/2, and 3/4 as fractions of the Y interval (this corresponds to $Y = 3/4$, $3/2$, and $9/4$). Solving for F from the cubic above, we get $F = 0.737$, 0.804, and 0.870, respectively. The resulting generalized elliptic equations are

$$X^{2.27} + Z^{2.27} = 1.25^{2.27}$$

$$X^{3.18} + Z^{3.18} = 1.50^{3.18}$$

$$X^{4.98} + Z^{4.98} = 1.75^{4.98}$$

The base of the exponent on the right side of each equation is the appropriate axis intercept as described by curves A and B of Fig. 58. Thus, the basic design constraints, concepts of generalized ellipses, and techniques of interpolation lead to the surface-describing equations as a function of Y. Although no single surface equation exists, we can derive any needed data from the presented definition technique, which is well suited to maintenance within a computer system. In effect, the computer generates surface data as

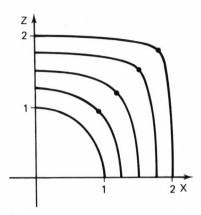

Figure 60 Surface cross sections projected into the *XZ* plane.

required. It does not store data as such, although that may be done if it is advantageous to do so. Figure 60 shows the *XZ* projections of the terminal and interpolated cross sections of our illustrative example.

In the illustration of this section, the boundary curves at the terminal points of the interval of interest are both mathematical equations. What if we had only a series of data points to represent the two boundary curves? What might we do? We might consider using an interpolation technique to establish *Z* for any *X* at both *Y* = 0 and *Y* = 3. This could be done by parametrically working with *Z* and *X* coordinates as independent functions of *θ*, where *θ* is measured from the origin of *XZ* (for a given *Y*). A line from that origin at an angle *θ* intersects the curve of interest. This affords equi-spacing of the parameter, which is an approximation of equal arc lengths on the curve for small increments of *θ*. This can be used for any set of coordinates or for mathematical equations like those for the illustration just discussed. Once *Z* and *X* are set up as functions of *θ*, then some method of interpolation between boundary curves (as a function of *Y*) must be established. The rule of interpolation might be (and usually would be) the same for any *θ*. One of the virtues of defining points on boundary curves at approximately equal arc lengths is its use in defining the *ruled surface*. A ruled surface is quite common in design. It is formed by connecting points on opposite boundaries with straight lines where the points of proportional arc lengths are connected. Wings of conventional aircraft are defined by this rule of connecting proportional

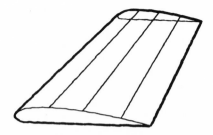

Figure 61 Ruled surface for an airplane wing section.

arc lengths of root and tip airfoils by straight lines (the teardrop shapes of wing cross sections are made by planes that are essentially parallel to the fuselage center line). When using a ruled-surface definition, intermediate cuts (interpolation) may be developed. This is shown in Fig. 61.

Of course, as many lines of connection as needed can be constructed for plotting or for use in analysis. The cross section for any intermediate location between boundary curves is determined by cutting the connecting lines with a plane, such as $Y = a$, that is, $0 \leqslant a \leqslant 3$ in the example depicted in Fig. 58.

Suppose as an illustration of developing a ruled surface, we return to the example at the beginning of this section. (We will remove the slope constants in the Y-direction for the example.) For both the circle at $Y = 0$ and the generalized ellipse at $Y = 3$, we will derive X and Z for ten increments of θ from $0°$ to $90°$ in $10°$ steps. The process of obtaining coordinates from the circle is quite well known and is simply $X = \cos \theta$ and $Y = \sin \theta$; $\theta = 0°, \ldots, 90°$. For the equation

$$X^{6.579} + Z^{6.579} = 2^{6.579}$$

the procedure is as follows: For a particular θ, the equation of the line through the origin is

$$(\tan \theta)X - Z = 0$$

Thus,

$$Z = (\tan \theta)X$$

and the equation becomes

$$X^{6.579} + (\tan \theta)^{6.579}X^{6.579} = 2^{6.579}$$

or

$$X^{6.579} = \frac{2^{6.579}}{1 + (\tan \theta)^{6.579}}$$

Solve for X by taking the 6.579th root and substituting it in the basic equation to derive Z. For small θ, Z is more easily found by noting that $Z = \tan(90° - \theta)$. The table below gives the resulting ten sets of coordinates for the two boundary curves as a function of θ.

	For $Y = 0$ $X^2 + Z^2 = 1$		For $Y = 3$ $X^{6.579} + Z^{6.579} = 2^{6.579}$	
θ	X	Z	X	Z
0°	1.000	0	2.000	0
10°	0.984	0.136	2.000	0.353
20°	0.940	0.342	2.000	0.728
30°	0.866	0.500	1.992	1.150
40°	0.766	0.643	1.918	1.610
50°	0.643	0.766	1.610	1.918
60°	0.500	0.866	1.150	1.992
70°	0.342	0.940	0.728	2.000
80°	0.136	0.984	0.353	2.000
90°	0	1.000	0	2.000

A ruled surface is derived by connecting the XZ coordinates for $Y = 0$ and $Y = 3$ for each increment of θ. To derive the corresponding coordinates for any Y in its interval, only linear interpolation is required. If, for example, we desire the curve for $Y = 3/4$ (which is $1/4$ of the distance from the initial boundary curve to the terminal boundary), then we need to determine the coordinates $1/4$ the distance from the first pair in the table to the second pair, and this is done for each θ. Thus, the "$1/4$" curve would be as follows:

Coordinates of Curve $1/4$ Distance from the Circle to the Generalized Elliptic Boundary		
θ	X	Z
0°	1.250	0
10°	1.240	0.190
20°	1.205	0.438
30°	1.147	0.661
40°	1.054	0.885
50°	0.885	1.054
60°	0.661	1.147
70°	0.438	1.205
80°	0.190	1.240
90°	0	1.250

Similar data sets can be developed for any θ. It is instructive to compare this ruled-surface interpolation with the previous interpolation process on the diagonals of generalized ellipses. The "1/4" curve then was

$$X^{2.27} + Z^{2.27} = 1.25^{2.27}$$

Suppose we select a few points for comparison. Based on the "1/4" curve for the two processes, we get:

θ	X	(Ruled) Z	(Original) Interpolation) Z
20°	1.204	0.438	0.410
30°	1.147	0.661	0.583
40°	1.054	0.885	0.758

This is a sample comparison. It is primarily a matter of the application type, the ease of automating, and the judgment of the analyst that determines the methodology that should be pursued in defining the surface intermediate to boundary curves. It is sufficient to say that there are many alternatives to this definition of which the two just presented are quite feasible and are in common use. We now turn to a practical example in design where the boundary designs, coupled with more varied processes of interpolation and extrapolation, represent a more sophisticated surface development process. In Section C-1 of this chapter, we discussed the development of boundary curves in space. In particular, Fig. 50 depicts the considerations in developing such curves to represent realistic airplane contours. (Here we will redesignate our coordinate system to conform with that discussion.) We stated how the describing curves might be developed, and we defined $Y = f(X)$ and $Z = g(X)$ as the resulting symbolic functions that produce the half-breadth contour. In addition, we will define $Z = h(X)$ as the resulting contour for the crown line in Fig. 50. No Y function is needed, since the crown is drawn exclusively in the XZ plane. Assuming, now, that these longitudinal boundary curves exist, we wish to define cross sections at critical locations along the airplane center line. From this data we wish to describe a meaningful process that uses the input definitions and design for the four boundary curves to create a surface. The situation is depicted in Fig. 62. Figure 62 is normal-

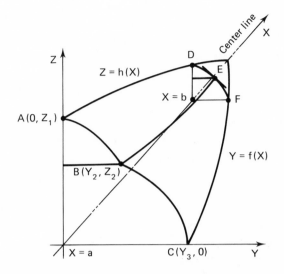

Figure 62 Input boundary data to develop a surface.

ized such that the X axis is the center line. Since the half-breadth curve is in the same horizontal plane as the center line, this makes $Z = g(X) = 0$ for all X. Now we have some additional design decisions. We want to create a cross section at $X = a$ because it is at this location where the constant cross section terminates (from front to rear). We also have a compartment and other equipment around which we wish to "wrap" the fuselage (at $X = a$) and at the same time keep the cross-sectional area at a minimum. (This is like the concept explained in Chap. II, Section G-3, Fig. 34.) It is our decision to do this with two blending elliptic arcs. Thus, at a display console, we elect to work in 2-D at $X = a$, and we input Z_2. We also input the location of the "crease" in the Y direction as indicated by Y_2. Both Z_1 and Y_3 are displayed, since $f(X)$ and $h(X)$ have been created and stored. Using the concepts developed in the section on ellipses, we must input two slopes—one for the top elliptic arc and another for the side elliptic arc. We are guided by the graphic display and the pictorial dialogue in inputing slopes consistent with the constraints as discussed in Chap. II, Section G-3. Further down the center line, we would like the crease to fade out—that is, the two arcs to become tangent. We specify this to transpire at $X = b$, and then the computer will provide a Z_1 from $h(X)$ and a Y_3 from $f(X)$ at $X = b$. The display will present those coordinates along with the local origin in a 2-D mode at $X = b$.

Now we may input Z_2 for $X = b$ but cannot, at this juncture, input Y_2. Again, along the lines presented earlier on elliptic blending, the computer must calculate and display a boundary Y_2 from formula (83) for $X = b$, which we denote as Y_2B $(X = b)$ where

$$0 \leqslant Y_2B(X = b) \leqslant Y_3(X = b)$$

Then we select a point at Z_2 that is within this range and denote it as Y_2D $(X = b)$. (The Y_2D means Y_2 designated.) The computer then displays the range of acceptable slopes according to the theory explained in Chap. II, Section G-3, and the designer specifies a slope in that range.

The process is now sufficiently developed for the computer to create the cross sections (first quadrant is all that is created, since symmetric properties account for the complete top section). Then it is a matter of interpolation to derive a cross section for any X between $X = a$ and $X = b$ and of extrapolation for any X greater than $X = b$.

Before proceeding, we would like to generalize on the cubic interpolation characterized by Fig. 59, that is, horizontal slopes at each end. This is done because of the extensive practical utility that this type of interpolation possesses, and it will be used in the present example. To develop this formula, we make two simple transformations. Whatever the range of the ordinate (vertical coordinate), we transform it to be from zero to unity. We do the same for the abscissa (horizontal coordinate), and we denote the two axes as Y and X, respectively. This is shown in Fig. 63.

The two points (0, 0) and (1, 1) with the two zero slopes give

$$Y = 3X^2 - 2X^3 \tag{99}$$

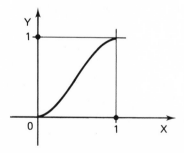

Figure 63 Cubic interpolation curve normalized to vary between zero and unity (in both directions).

Should we have the complementary curve of that shown in Fig. 63, that is, from (0, 1) to (1, 0) with zero slopes, then

$$Y = 1 - (3X^2 - 2X^3) \qquad (99a)$$

To illustrate Eq. (99), suppose Y varies from 90 to 120 while X varies from 3 to 5, and we wish to find Y at $X = 3\frac{1}{2}$. Normalized, $X = (3\frac{1}{2} - 3)/(5 - 3) = 1/4$. Then,

$$Y = \left[3\left(\frac{1}{4} \right)^2 - 2\left(\frac{1}{4} \right)^3 \right] 30 + 90$$

$$Y = 94.69$$

Now let us return to the example depicted by Fig. 61. Consider interpolation first. We let X_i be any X in the interval from $X = a$ to $X = b$. We already know Z_1 and Y_3 from given input. We want to find Y_2, Z_2, and the two slopes at X_i such that two elliptic arcs may be defined at any X_i. Since $X = a$ is at the termination of a constant cross section (given), then both Y_2 and Z_2 would not be varying with X up to position $X = a$. However, Fig. 62 reveals that both are decreasing with X to $X = b$ and beyond. Thus, a plot of either Y_2 or Z_2 versus X in the interval would have a zero slope at $X = a$ but a negative slope at $X = b$. But it is a very reasonable assumption that the ratios of Z_2 to Z_1 and of Y_2 to Y_3 remain constant from $X = b$ to the rear terminus. Therefore, we assert that both Z_2/Z_1 and Y_2/Y_3 vary (Y_2 is the designated Y_2 that we called Y_2D earlier) along a cubic path and have zero slopes at $X = a$ and at $X = b$. Thus, either formula (99) or (99a) may be applied. Let us work with Z_2/Z_1 realizing that an analogous process applies for Y_2/Y_3. We select X_i and compute

$$X_i^* = \frac{X_i - a}{b - a} \qquad (100)$$

where X^* is the fraction of the interval. Then,

$$\frac{Z_2}{Z_1}(X_i) = [3X_i^{*2} - 2X_i^{*3}] \left[\frac{Z_2}{Z_1}(X = b) - \frac{Z_2}{Z_1}(X = a) \right]$$

$$+ \frac{Z_2}{Z_1}(X = a) \qquad (101)$$

All the data for formula (101) is known since X_i^* is found by formula (100) and

$$\frac{Z_2}{Z_1} (X = b)$$

and

$$\frac{Z_2}{Z_1} (X = a)$$

are found from the cross section data input as derived at $X = a$ and $X = b$; that is, Z_2 is input at a and b and Z_1 is taken from $Z_1 = h(X)$—input from the longitudinal development.

Let the entire right side of formula (101) be denoted by K. Then we get the desired $Z_2(X_i)$ by

$$Z_2(X_i) = Z_1(X_i)K \tag{102}$$

(Again, $Z_1(X_i)$ is found from $Z_1 = h(X)$.)

Similarly, formulas (99) through (102) yield Y_2 (we need only substitute Y_2/Y_3 for Z_2/Z_1, Y_3 for Z_1, and $Y_3 = f(X)$ for $Z_1 = h(X)$ in the appropriate places). By this procedure, we derive Y_2 and Z_2 for X_i. We must still derive the two slopes at point B (Fig. 62) to have the needed data from which the two elliptic arcs can be constituted. We know the two slopes at $X = a$ and the common slope at $X = b$. The question is, how do we interpolate? One thing must be understood, and that is the fact that we cannot interpolate on slopes, per se. A vertical line has an infinite slope; so if we were to try to find a line intermediate to two lines, one of which is vertical, we could not find an intermediate slope between a finite slope and an infinite slope. Mathematically, we might perform slope interpolation when both lines have the same slope sign and neither is vertical, but the property just discussed would usually make such an interpolation unrealistic. Slopes are not proportional to angles, and we are usually more concerned with the angles that the lines make. Therefore, we suggest that interpolation on line position or orientation be based on angles rather than slopes. Thus,

$$\alpha = \tan^{-1}|Z'| = \tan^{-1} (\text{slope at point } B) \tag{103}$$

Generally, we can work with absolute values of slopes as in Eq. (103) as long as we are careful to set up the correct sign after interpolation. Now we use Eq. (104) to set up more specific definitions. The reader is asked to note certain notation changes at this point that are essential to understand the following formulas. Up to this point, the subscript *1* has pertained to coordinates at the top (as Y_1 and Z_1 of Fig. 62). Similarly the subscript *2* has pertained to the coordinates at the crease where two arcs are joined, and the subscript *3* has pertained to coordinates at the side. This will remain the case when we use coordinates. However, we are going into a more detailed description of slopes, angles, and ratios that pertain to the top arc and the side arc at their junction. Therefore, the subscript *1* for Z_1', $Z_{1, L}'$, α_1, $\alpha_{1, L}$, and R_1 will pertain to the top arc at the crease—not the top point of the cross section. Similarly, the subscript *2* will pertain to the side arc at the crease. With these concepts in mind, we continue.

$$\alpha_1(X = a) = \tan^{-1}|Z_1'(X = a)| \qquad (104)$$

(the angle made by the tangent line for the top arc at the crease. Z_1' at $X = a$ is input from earlier development).

$$\alpha_2(X = a) = \tan^{-1}|Z_2'(X = a)|$$

(the same formula for the side arc). $Z_2'(X = a)$ was also input earlier.

$$\alpha_{1, L}(X = a) = \tan^{-1}|Z_{1, L}'(X = a)|$$

(the angle made by the limiting or boundary tangent line for the top arc). Having previously input and derived Z_1, Z_2, and Y_2, then $Z_{1, L}'(X = a)$ is given by (from Chap. II, Section G-3)

$$Z_{1, L}'(X = a) = 2\,\frac{Z_2(X = a) - Z_1(X = a)}{Y_2(X = a)}$$

Similarly for the side arc, $\qquad\qquad\qquad\qquad\qquad\qquad\qquad (104a)$

$$\alpha_{2, L}(X = a) = \tan^{-1}|Z_{2, L}'(X = a)|$$

where

$$Z_{2, L}'(X = a) = \frac{1}{2}\,\frac{Z_2(X = a)}{Y_2(X = a) - Y_3(X = a)}$$

Thus, all angles at $X = a$ for the slopes may be derived from data developed when the cross section was created at $X = a$.

At $X = b$ there is a single common slope that is input according to previous descriptions. There are two limiting slopes at $X = b$. These limiting slopes (though we could not use both simultaneously at $X = b$ and have a blending of the two arcs) are needed for interpolation to ensure that angles do not exceed limits from which arcs are constructed. This is a very subtle concept and usually cannot be comprehended without considerable thought and study. Nevertheless, the principle is correct and the formulas we derive may be used whether or not their derivation is comprehended. To continue,

$$\alpha_1(X = b) = \alpha_2(X = b) = \tan^{-1}|Z'(X = b)|$$

where $Z'(X = b)$ is input in the development of the $X = b$ cross section,

$$\alpha_{1,L}(X = b) = \tan^{-1}|Z'_{1,L}(X = b)|$$

and   (104b)

$$\alpha_{2,L}(X = b) = \tan^{-1}|Z'_{2,L}(X = b)|$$

$Z'_{1,L}$ and $Z'_{2,L}$ for $X = b$ are computed analogously to the way those limits were computed at $X = a$.

Having derived all the angles of concern, we are now ready to derive the interpolated angles for X_i, the value of X between $X = a$ and $X = b$. We make the following definitions to simplify the computation and the data "bookkeeping."

$$R_1(a) = \frac{\alpha_1(X = a) - \alpha_{1,L}(X = a)}{\pi/2 - \alpha_{1,L}(X = a)}$$

$$R_2(a) = \frac{\alpha_2(X = a)}{\alpha_{2,L}(X = a)}$$

$$R_1(b) = \frac{\alpha_1(X = b) - \alpha_{1,L}(X = b)}{\pi/2 - \alpha_{1,L}(X = b)}$$

and

$$R_2(b) = \frac{\alpha_2(X = b)}{\alpha_{2,L}(X = b)}$$

Each of the R's are fractions less than unity, which constrain the angles from falling outside the limiting values. For the top arc, we interpolate on the ratios (the R's) and select the special type of cubic as exemplified by formulas (99) through (102). Using these formulas, we derive $R_1(i)$ between $R_1(a)$ and $R_1(b)$. Then we get the angle at X_i by

$$\alpha_1(X = X_i) = R_1(i)\left[\frac{\pi}{2} - \alpha_{1,L}(X = X_i)\right] + \alpha_{1,L}(X = X_i)$$

where $\alpha_{1,L}(X = X_i)$ is obtained from given and derived data in a procedure like that used to find $\alpha_{1,L}(X = a)$ and $\alpha_{1,L}(X = b)$.

Finally we get the slope $m_1(X = X_i)$ by

$$m_1(X = X_i) = \pm\tan\alpha_1(X = X_i)$$

—the sign depending on the sign of the initial slopes at $X = a$ and $X = b$, negative for the first quadrant as in Fig. 62. For the side arc, we again use formulas (99) through (102) to interpolate between $R_2(a)$ and $R_2(b)$ to get $R_2(i)$. Then, similar to the top arc,

$$\alpha_2(X = X_i) = R_2(i)\alpha_{2,L}(X = X_i)$$

(There is a slight difference between this formula and that for the top arc because this angle is between zero and a limit rather than $\pi/2$ and a limit.)

Having discussed interpolation, we now turn to extrapolation. We would like to use existing definitions and derivations to define a cross section for $X \geqslant b$, which we will denote as X_j. Even though the two arcs are blended for all X_j, we must still derive Y_2 and Z_2 where this blend occurs and an appropriate slope such that the arc equations may be duly derived. Proper extrapolation technique will ensure that the surface, as generated by cross section definition for all X will be continuous across $X = b$. We recall earlier in the discussion that we computed $Y_2B(X = b)$ from formula (83) to get the minimum Y_2 that could be input and still ensure two-arc blending (tangency). Then the Y_2 we input between that value and $Y_3(X = b)$ was denoted as $Y_2D(X = b)$. Now we wish to find Y_2 beyond $X = b$, which is similar to the design depicted in Fig. 62 and which can be automatically derived when X_j is specified. We must take care to keep this Y_2, which we denote as $Y_2(X = X_j)$, within the Y_2B and Y_3 limits at $X = X_j$ to ensure a solution. Looking at Fig. 62 and noting how the geometry converges as X increases, we make a design decision that

$$R_j(Y_2) = \frac{Y_2D(X = X_j) - Y_2B(X = X_j)}{Y_3(X = X_j) - Y_2B(X = X_j)}$$

should be constant. Most of the input for $R_j(Y_2)$ already exists so that we can solve for Y_2D. We have $Y_3 = f(X)$ and $Z_1 = h(X)$. To find Y_2B we also need $Z_2(X = X_j)$. Using the same rationale as for the Y_2's, we wish to preserve the ratio that exists at $X = b$, that is, $Z_2(X = b)/Z_1(X = b)$. Thus, we find $Z_1(X = X_j)$ from $Z_1 = h(X)$ and multiply it by this ratio to derive $Z_2(X = X_j)$. Then the Z_1, Z_2, and Y_3 at that location suffice to derive $Y_2B(X = X_j)$ from formula (83). How do we find the value of $R_j Y_2$? We want it to be the same as it is at $X = b$, $R_b Y_2$. We have all the data for $R_b Y_2$—namely Y_2D, Y_2B, and Y_3 at $X = b$. Therefore,

$$\frac{Y_2D - Y_2B}{Y_3 - Y_2B} = R_j Y_2 = R_b Y_2$$

Now we have developed all components of the equation except Y_2D; so we solve the equation for it. Now we have Y_2 and Z_2 at $X = X_j$. To have sufficient input to obtain the two elliptic arcs, we still must extrapolate to get the common slope of the two curves. Recall that we input $Z'(X = b)$, which is the slope at (Y_2, Z_2) at $(X = b)$. We had to input a value between two limiting slopes as discussed in Chap. II, Section G-3. Working with the angles equivalent to those derived from $\alpha = \tan^{-1}|Z'|$ and as indicated in the interpolation discussion, we wish to keep the same ratios in extrapolation. This concept may be more easily understood with Fig. 64. The designated (or derived) angle, α_D, must be in the band α_1 to α_2, which are the limiting angles that produce a solution. This is required regardless of the location of X, which is where we seek a solution.

We can find the ratio $R(\alpha)$ at $X = b$ by

$$R(\alpha) = \frac{\alpha_D - \alpha_1}{\alpha_2 - \alpha_1} (X = b)$$

because all components of the ratio have been derived or input. We maintain this ratio at $X = X_j$. Thus,

$$R(\alpha) = \frac{\alpha_D - \alpha_1}{\alpha_2 - \alpha_1} (X = X_j)$$

We can derive α_1 and α_2 at $X = X_j$ from the Y_2 and Z_2 (just derived) and the known functions of Z_1 and Y_3. Therefore, we need only to solve the above equation for α_D $(X = X_j)$ and convert it to a slope by calculating $\tan \alpha_D$. The $(Y_2 Z_2)$ point and the slope are the

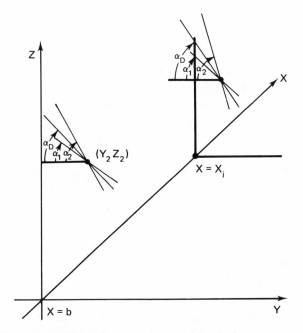

Figure 64 Diagram for the extrapolation formula for the
common tangent line.

new data needed to add to the existing data in order to derive the
two elliptic arcs.

With the slopes and (Y_2, Z_2) derived automatically at X_i or X_j,
the procedures for elliptic arc (or other geometric) development can
be implemented to give the desired cross section at any X. The total
process is the systematic development of boundary curves including
cross section development at key locations and the use of programmed interpolation and extrapolation procedures that find first
the basic data and then the appropriate curves. No single step is, in
itself, very complicated. What is difficult is learning the many and
varied options; keeping off-line input data, on-line input, and output in proper perspective; and being able to weave the processes
together without error to achieve the desired application objectives.
This whole process would become impractical without computer
graphic dialogue as an integral feature of the system development.

There are many options to be followed in surface development.
We have not enumerated the equations in this chapter as we did in
Chap. II because we have described example processes. The equa-

tions and formulas herein may not be appropriate, without modification, to other applications. For example, the blending of ellipses is only one way to wrap a surface envelope around a perceived shape. Thus, the descriptions of this section are intended primarily to exemplify concepts and are, by no means, all encompassing.

To keep the various parameters and variables properly delineated is a notational nightmare that may block out some understanding of fundamentals. In some cases there is more notation than needed from a purist point of view, but it is done with the intention of helping to keep things straight—like $Y_2D(X = X_j)$. However, it is hoped that the reader will be able to discern the essential concepts that we have attempted to convey.

E. SURFACES FORMED FROM FOUR GENERAL SPACE CURVES

In the previous section we discussed the formation of surfaces from space curves, where two curves were the primary influencing functions and, therefore, interpolation was performed between those curves. Now we will consider a more general system of four boundary curves, where each curve is essentially equally important in influencing intermediate or interpolated data. This type of approach has been well treated by S. A. Coons.[1] We will expound on some of the more basic concepts as set forth by Coons.

A space curve may be expressed either as a series of data points or as a mathematical formulation. Additional data can be derived as needed by curve fitting (interpolation) in the former case or by direct equation evaluation in the latter case. (One or two space curves may be degenerate, that is, composed of a single point.) A set of such curves is shown in Fig. 65.

The four space curves are identified in each view as A, B, C, and D. The process of developing surface contours is a systematic definition of contours that connect proportional arc lengths of opposite boundaries. These boundary curves are taken from an actual design problem in determining the surface connection between a wing and a fuselage. Their development is an interesting

[1]S. A. Coons, "Surfaces for Computer-Aided Design of Space Forms," M. I. T. Project MAC, MAC-TR-41 (also AD 663 504), June 1967.

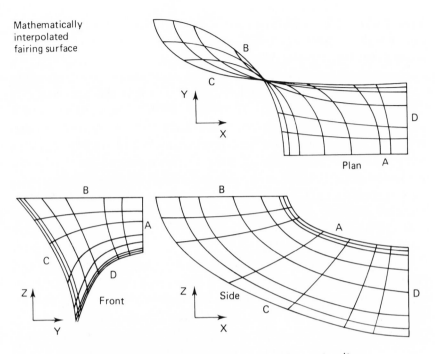

Mathematically
interpolated
fairing surface

B

C

Y

X

D

Plan   A

B

A

C

D

Z

Y

Front

B

A

Z

Side

C

X

D

Figure 65 Surface fit to boundary curves for a wing/fuse-
lage airplane fairing.

process in itself, but the discussion here will deal with the development of the surface from the boundaries. This begins by noting that
each space curve can be represented by parametric relationships
between each variable and arc length. It should be noted that the
arc length between two points on a curve is generally approximated
by the chord length between the points. Since a vector (chord
length) is composed of X, Y, and Z components, it follows that no
component can be larger than the arc length to which it relates.
Therefore, a plot of any component versus arc length will never
develop a slope greater than unity. Thus, the breaking up of a space
curve into three parametric curves of X, Y, and Z versus arc length
insures a more tractable functional relationship for each variable.
The way this is done in practical terms is that each variable is
independently related to arc length. The total span of each arc is
normalized (a mapping, change of variable process) to unity and
each arc-length variable is arbitrarily called U for two curves and V
for the other two. This is more easily understood with the use of an
example that we extract from the boundary curve data of Fig. 65.

We will arbitrarily work with the Z coordinates of the four curves, although the same process applies for X and Y. Figure 66 depicts Z versus U and V.

Figure 66 contains a great deal of information; so it warrants careful study and additional explanation. What we are doing is mapping the Z coordinates such that we will be able to set up meaningful interpolation procedures. Take curve C of Fig. 65, for example. We extract only the Z coordinates, and we make a 2-D plot of Z versus arc length of curve C. We quite arbitrarily denote the arc-length variation as U and, regardless of actual length, it is mapped onto a unit length. The curve opposite in space to C is A. We also want to make a 2-D plot of Z from curve A, but we must locate it apart from the first plot. Thus, Z of curve C is plotted versus U at $V = 0$, and Z of curve A is plotted versus U at $V = 1$. Symbolically, the two plots are represented by $Z(U, 0)$ and $Z(U, 1)$, respectively. Similarly, the Z coordinates of B and D are represented by $Z(0, V)$ and $Z(1, V)$, respectively. The Z coordinates at the points of intersection are $Z(0, 0)$, which is the Z coordinate for $U = 0$ and $V = 0$; $Z(1, 0)$, which is the coordinate for $U = 1$ and $V = 0$; and so on. Now we have constructed four boundary curves for Z from which we can derive a surface for Z for ranges of U and V between zero and unity. Let us suppose that we wish to derive a Z for any U, V. We select $U = U_1$ and $V = V_1$. U_1 is a certain fraction of unity and V_1 is another fraction of unity. Those fractions on the UV axes of Fig. 66 lead to Z coordinates as represented by P, Q, R, and S. Those four points along with the

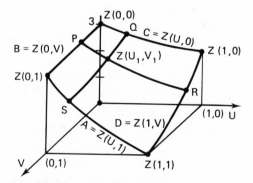

Figure 66 Z-coordinates of Fig. 65 mapped over the unit square in the UV plane.

corner points are the basic data from which $Z(U_1, V_1)$ (a point on the surface at $U_1 V_1$) is found by an interpolation process. Quite naturally, the closer $U_1 V_1$ is to the various coordinates P, Q, R, S, and the corners, the more weight they should carry in interpolation. In order to insure a continuous transition across boundaries for adjacent surfaces, S. A. Coons shows that the interpolation formulas (or blending functions) should be of the form described by formulas (99) and (99a) in the previous section, that is, $3U^2 - 2U^3$ and its complement $1 - (3U^2 - 2U^3)$ when interpolating in the U direction. The same form applies in the V direction with U replaced by V.

For brevity we define

$$B(U) = 1 - (3U^2 - 2U^3)$$
$$1 - B(U) = 3U^2 - 2U^3$$
$$B(V) = 1 - (3V^2 - 2V^3)$$
$$1 - B(V) = 3V^2 - 2V^3$$

We note that $B(U) = 1$ for $U = 0$, and $B(U) = 0$ for $U = 1$. We use $B(U)$ notation instead of some other designation because B stands for *blending* and is easier to work with in the formulas that follow. Using the foregoing definitions, we might interpolate to find $Z(U_1, V_1)$ as follows:

$$\begin{aligned}
Z(U_1, V_1) = {} & P \cdot B(U_1) + R \cdot [1 - B(U_1)] + Q \cdot B(V_1) \\
& + S \cdot [1 - B(V_1)] - Z(0, 0)B(U_1)B(V_1) \\
& - Z(1, 0)[1 - B(U_1)]B(V_1) \\
& - Z(0, 1)B(U_1)[1 - B(V_1)] \\
& - Z(1, 1)[1 - B(U_1)][1 - B(V_1)] \qquad (105)
\end{aligned}$$

Formula (105) looks, at first glance, like a complicated relationship; but the B functions are nothing more or less than weighting coefficients to blend the eight Z coordinates, P, Q, R, S, $Z(0, 0)$, $Z(1, 0)$, $Z(0, 1)$, and $Z(1, 1)$.

To illustrate the use of formula (105), suppose we can derive the eight values of Z from known boundary curves and the specification of the fractions U_1 and V_1. Suppose that $U_1 = 1/3$ and

$V_1 = 1/4$, for example. Then,

$$B(U) = \frac{20}{27}, \qquad 1 - B(U) = \frac{7}{27}$$

$$B(V) = \frac{27}{32}, \qquad 1 - B(V) = \frac{5}{32}$$

Therefore from formula (105),

$$Z\left(\frac{1}{3}, \frac{1}{4}\right) = P \cdot \left(\frac{20}{27}\right) + R \cdot \left(\frac{7}{27}\right) + Q \cdot \left(\frac{27}{32}\right)$$

$$+ S \cdot \left(\frac{5}{32}\right) - Z(0, 0) \cdot \left(\frac{20}{27} \cdot \frac{27}{32}\right)$$

$$- Z(1, 0) \cdot \left(\frac{7}{27} \cdot \frac{27}{32}\right) - Z(0, 1) \cdot \left(\frac{20}{27} \cdot \frac{5}{32}\right)$$

$$- Z(1, 1) \cdot \left(\frac{7}{27} \cdot \frac{5}{32}\right) \tag{106}$$

As an exercise, add all of the positive and negative coefficients together. You should find that they sum to unity, which they must in order to have a meaningful and useful interpolation process. If we wanted to test formula (105) for $U_1 = 0$ and $V_1 = 0$, we would note that both P and Q become $Z(0, 0)$ at $U_1 = V_1 = 0$. We also note that $B(0) = 1$ and $1 - B(0) = 0$. Thus, formula (105) becomes

$$Z(0, 0) = Z\overset{P}{(0, 0)} \cdot (1) + R \cdot (0) + Z\overset{Q}{(0, 0)} \cdot (1)$$

$$+ S \cdot (0) - Z(0, 0) \cdot (1) \cdot (1) - Z(1, 0) \cdot (0) \cdot (1)$$

$$- Z(0, 1) \cdot (1) \cdot (0) - Z(1, 1) \cdot (0) \cdot (0)$$

or

$$Z(0, 0) = Z(0, 0)$$

which verifies that the formula works at $(0, 0)$. Now we return to the evaluation of $Z(1/3, 1/4)$ as exhibited by formula (106). We will go to the data from which Figs. 65 and 66 were drawn. (The data are normalized to simple values to aid in this illustration. They are not the actual dimensions, of course.) We have

$$P = 3, \qquad Q = 2.34, \qquad R = 1.50, \qquad S = 1.10$$

$$Z(0, 0) = 3, \qquad Z(1, 0) = 2, \qquad Z(0, 1) = 3, \qquad Z(1, 1) = 0$$

Thus, formula (106) gives

$$Z\left(\frac{1}{3},\frac{1}{4}\right) = 3\left(\frac{20}{27}\right) + 1.5\left(\frac{7}{27}\right) + 2.34\left(\frac{27}{32}\right) + 1.10\left(\frac{5}{32}\right)$$

$$- 3\left(\frac{20}{27}\right)\left(\frac{27}{32}\right) - 2\left(\frac{7}{27}\right)\left(\frac{27}{32}\right)$$

$$- 3\left(\frac{20}{27}\right)\left(\frac{5}{32}\right) - 0\left(\frac{7}{27}\right)\left(\frac{5}{32}\right)$$

$$Z\left(\frac{1}{3},\frac{1}{4}\right) = 2.1$$

Similarly, a Z coordinate on the surface can be derived for any (U, V) in the range of definition. As for Z, X and Y are derived by an identical process; so by varying U and V, the variables X, Y, and Z are generated. Formula (105) may be written to apply to any U, V for each variable. Thus, $F(U, V)$ is used to denote each of the X, Y, and Z variables. The general formula is, therefore,

$$F(U, V) = F(0, V)B(U) + F(1, V)\big[1 - B(U)\big]$$

$$+ F(U, 0)B(V) + F(U, 1)\big[1 - B(V)\big]$$

$$- F(0, 0)B(U)B(V) - F(0, 1)B(U)\big[1 - B(V)\big]$$

$$- F(1, 0)\big[1 - B(U)\big]B(V)$$

$$- F(1, 1)\big[1 - B(U)\big]\big[1 - B(V)\big] \tag{107}$$

In formula (107), the F functions on the right are boundary curves for each variable. Such curves might be data or actual mathematical expressions from which the values P, Q, R, and S are derived for particular values of U and V. If the curve is a cubic, then a cubic function of U (for example) multiplied by the cubic blending function of V produces what is commonly known as a *bicubic*, and there will be a bicubic expression for each of the four principal terms. There is good reason to use the same type of interpolation form as is used for the boundaries, although it is not an absolute requirement. Many times there are practical difficulties in imposing this requirement.

The natural question arises at this point as to the physical properties and goodness of fit to aesthetic preferences. How well would this technique fit or match a known control surface? For example, we know the equation of the surface of a sphere. If we work only with the boundary curves and use formula (107), how well do we develop the spherical shape? To demonstrate this, we consider the circular boundaries of an octant of a sphere of unit radius as depicted in Fig. 67. (This example also illustrates the process of creating a surface according to the principles of this section.)

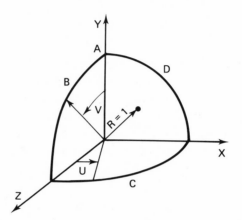

Figure 67 Control spherical surface to be approximated by interpolation between boundary curves.

There are four boundary curves—curve A, which is the single point at the top (degenerate curve), curve B, which is the circular arc at the left, curve C, which is the lower arc, and curve D, which is the arc on the right. We let U and V be the two angles depicted in the figure. The four boundary equations, $F(U, V)$, for each variable are as follows:

Curve	X	Y	Z
A	0	1	0
B	0	$\cos \frac{\pi}{2} V$	$\sin \frac{\pi}{2} V$
C	$\sin \frac{\pi}{2} U$	0	$\cos \frac{\pi}{2} U$
D	$\sin \frac{\pi}{2} V$	$\cos \frac{\pi}{2} V$	0

Then from (107),

$$F(U, V) = X(U, V) = 0\,B(V) + \left[\sin\,\frac{\pi}{2}\,U\right]\left[1 - B(V)\right]$$

$$+ 0\,B(U) + \left[\sin\,\frac{\pi}{2}\,V\right]\left[1 - B(U)\right]$$

$$- 0\,B(U)B(V) - 0\,B(V)\left[1 - B(U)\right]$$

$$- 0\,B(U)\left[1 - B(V)\right]$$

$$- 1\left[1 - B(U)\right]\left[1 - B(V)\right]$$

or

$$X(U, V) = \left[\sin\,\frac{\pi}{2}\,U\right]\left[1 - B(V)\right] + \left[\sin\,\frac{\pi}{2}\,V\right]\left[1 - B(U)\right]$$

$$- \left[1 - B(U)\right]\left[1 - B(V)\right]$$

Similarly, $Y(U, V)$ simplifies to

$$Y(U, V) = \cos\,\frac{\pi}{2}\,V \tag{108}$$

and

$$Z(U, V) = \left[\cos\,\frac{\pi}{2}\,U\right]\left[1 - B(V)\right] + \left[\sin\,\frac{\pi}{2}\,V\right]B(U)$$

$$- B(U)\left[1 - B(V)\right]$$

The three formulas for X, Y, and Z as functions of U and V make up formula (108). We merely need to evaluate $B(U)$, $B(V)$, and so on, along with $\sin\,\frac{\pi}{2}\,U$, $\sin\,\frac{\pi}{2}\,V$, $\cos\,\frac{\pi}{2}\,U$, and $\cos\,\frac{\pi}{2}\,V$ for values of U and V between 0 and 1. The distance of a point generated by this method from the origin will be

$$d = (X^2 + Y^2 + Z^2)^{1/2}$$

to be compared to $R = 1$.

We select $U = 1/3$, $1/2$, and $2/3$; and $V = 1/3$, $1/2$, and $2/3$. From trigonometry,

$$\sin\,\frac{1}{3}\left(\frac{\pi}{2}\right) = 0.50 \qquad \cos\,\frac{1}{3}\left(\frac{\pi}{2}\right) = 0.87$$

$$\sin\,\frac{1}{2}\left(\frac{\pi}{2}\right) = 0.71 \qquad \cos\,\frac{1}{2}\left(\frac{\pi}{2}\right) = 0.71$$

$$\sin\,\frac{2}{3}\left(\frac{\pi}{2}\right) = 0.87 \qquad \cos\,\frac{2}{3}\left(\frac{\pi}{2}\right) = 0.50$$

Also

$$B\left(\frac{1}{3}\right) = 1 - \left[3\left(\frac{1}{3}\right)^2 - 2\left(\frac{1}{3}\right)^3\right] = \frac{20}{27}$$

$$B\left(\frac{1}{2}\right) = 1 - \left[3\left(\frac{1}{2}\right)^2 - 2\left(\frac{1}{2}\right)^3\right] = \frac{1}{2}$$

$$B\left(\frac{2}{3}\right) = \frac{7}{27}$$

The following tables result from the substitution of these values into formula (108).

X

U \ V	0	1/3	1/2	2/3	1
0	0	0	0	0	0
1/3	0	0.19	0.30	0.40	0.50
1/2	0	0.30	0.46	0.59	0.71
2/3	0	0.40	0.59	0.74	0.87
1	0	0.50	0.71	0.87	1.00

Y

U \ V	0	1/3	1/2	2/3	1
0	1.00	→			
1/3	0.87	→			
1/2	0.71	→	For all U		
2/3	0.50	→			
1	0	→			

Z

U \ V	0	1/3	1/2	2/3	1
0	0	0	0	0	0
1/3	0.50	0.40	0.30	0.19	0
1/2	0.71	0.59	0.46	0.30	0
2/3	0.87	0.74	0.59	0.40	0
1	1.00	0.87	0.71	0.50	0

From these three tables we can compute the distance of each surface grid point from the origin by $d = (X^2 + Y^2 + Z^2)^{1/2}$ for each (U, V). Thus, we have

$\frac{U}{V}$	0	1/3	1/2	2/3	1
0	1	1	1	1	1
1/3	1	0.972	0.965	0.972	1
1/2	1	0.968^+	0.962	0.968^+	1
2/3	1	0.979	0.973	0.979	1
1	1	1	1	1	1

Figure 68 Table of interpolated values of d using the conventional basic Coons technique.

If we study the table of Fig. 68, we see the interpolated coordinates converted to distances from the origin for a network of U, V values. We would like all of them to be equal to unity to conform to the sphere. In reality, we derive a flattened spherical shape with the surface falling short of unity from 2% to a maximum of almost 4% over a sizeable section. For many applications, that margin of error might be perfectly acceptable. However, we may ask what might be done to improve the interpolation accuracy in developing a truly spherical shape *without significantly increasing the complexity* of the system for a graphic user? We want to reemphasize the fact that we have no difficulty in representing a spherical surface precisely, but our objective is to make one test of the generality of a technique under controlled conditions. The next section addresses the question of improved interpolation.

F. VARIABLE INTERPOLATION FOR FLEXIBLE SURFACE DEVELOPMENT

Mr. Coons has developed a *correction* surface which can be adjoined to that of formula (107) to get a much better fit. This requires the input of cross derivatives at the corners where the boundary curves intersect. It is difficult for most people to get a "feel" for the relationship between such input and the resulting surface. There is little easily visualized geometric significance to the numerical input. Also, several independent derivatives must be input which make it difficult to perform efficient trial and error practices to fine tune the input data. It is for these practical considerations that we wish to present an alternative approach to modifying the surface—an approach "made to order" for computer graphics. To explain this approach, it should be noted that the weighting functions, $B(V)$ and $1 - B(V)$, are complementary and symmetrical functions; that is, opposite boundary curves are weighted strictly according to their distances in the V direction. No special increased or decreased weight is given because of the relative length of the boundary curves. Thus, in Fig. 67, each point of curve A (a curve of a single point) has the same effect as the proportional point on curve C. We believe that this condition is too restrictive. The cubic blending formulas (99) and (99a) discussed previously are depicted in Fig. 69.

These cubic curves determine the relative weight of boundary curves $F(0, U)$ and $F(1, U)$. The value $B(V)$ applies to the first curve and $1 - B(V)$ applies to the second—regardless of the lengths of the two curves. We now introduce the concept of a variable blending function. Suppose in Fig. (69) we change the point at $(1/2, 1/2)$ to $(1/2, K)$. This constraint along with the two end points and slopes give a total of five constraints and changes our cubic to a quartic. This leads to $B'(V)$ and $1 - B'(V)$ where

$$B'(V) = (16K - 8)V^4 + (18 - 32K)V^3 + (16K - 11)V^2 \quad (109)$$

(For $K = 1/2$, $B'(V)$ reduces to $B(V)$—which it should.) Formula (109) is extraordinarily easy to implement. The parameter K can be varied between zero and unity by inputing trial values by keyboard

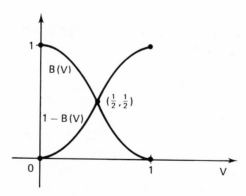

Figure 69   Cubic blending functions for Coons surfaces.

or by slowly turning a knob. The idea of turning a knob is well suited to graphics. By that means, the operator changes the blending function between two boundary curves and immediately sees the result. There could be two such parameters, say K and L, such that the blending might be tuned in both U and V directions. At most there would be two parameters that can be varied on-line at a console. Either or both parameters, when set equal to $1/2$, cause the blending equation to convert back to the basic cubic forms. Thus, formula (109) should prove to be very useful for practical surface-fitting applications.

Let's see how much formula (109) helps improve the fit of the spherical example of the previous section. To do this, we will use formula (109) to interpolate in the V direction but will continue to use the basic cubic of formulas (99) and (99a) when interpolating in the U direction. (The latter decision is made because interpolation in the U direction is between boundaries of equal length.) We will not compute each element of the three tables that leads to the table of Fig. 68. However, we will use selected values of K (including $K = 0.5$, which has already been computed in the previous tables) and compute the middle column; that is, $(U, V) = (1/2, V)$ for $K = 0.2, 0.3, 0.4, 0.6, 0.7,$ and 0.8. We want to see how the change of K affects the surface in our attempt to match the unit spherical segment.

$$X$$

$(U=1/2)$ $K =$ V	0.2	0.3	0.4	Original Column, 0.5	0.6	0.7	0.8
0	0	0	0	0	0	0	0
1/3	0.35	0.335	0.32	0.30	0.29	0.27	0.25
1/2	0.52	0.50	0.48	0.46	0.44	0.42	0.40
2/3	0.635	0.62	0.61	0.59	0.57	0.555	0.54
1	0.71	0.71	0.71	0.71	0.71	0.71	0.71

This table is computed using selected K values, interpolation by Eq. (109) in the V direction, and the surface generation Eq. (108). From Eq. (108), we observe that the Y coordinates are unaffected by K and that Z versus K (for $U = 1/2$) yields the exact same table just generated for X. Thus, distances, d, are recomputed by using the X from the above table for both X and Z and by using the previous Y's that were generated by $Y = \cos \pi/2 \ V$. This leads to the table showing surface distances of an approximate spherical shape as tested for a variation in the interpolation parameter, K, through the middle of the surface, that is, $U = 1/2$.

d

(U=1/2) K V	0.2	0.3	0.4	Original Data, 0.5	0.6	0.7	0.8
0	1	1	1	1	1	1	1
1/3	0.998	0.988	0.976	0.965	0.958	0.947	0.935
1/2	1.020	1.000	0.980	0.962	0.940	0.924	0.906
2/3	1.028	1.009	0.996	0.973	0.948	0.933	0.940
1	1	1	1	1	1	1	1

    The object in developing this table is to find the value of K that results in the best fit to a spherical test surface in the interior ($U = 1/2$) of the surface. The original basic cubic interpolation is represented by $K = 0.5$. We observe a maximum error (between d and unity) of about 4% in that column. We observe that for $K = 0.2$ there is an excellent match for a considerable range of V with a maximum error of less than 3%. For $K = 0.3$, the maximum error in making the match is just over 1%. Although we have not computed the error for all V, a plot of error versus V and each K would reveal that we are close to the absolute maximum. It is also interesting to note that the plot of d versus V for any K gives a slight oscillation. Often we are concerned only with the closeness of the fit, but some applications, such as those that may involve fluid or gas flow over a surface, may be concerned with oscillations. As always, the nature of the application influences the choice of surface-generation options including whether or not variable interpolation is warranted. We will now use the *best* K, which we select as 0.3, and recompute d versus U and V just as we did for Fig. 68. We have the column for $U = 1/2$ and need only compute the additional columns for $U = 1/3$ and $U = 2/3$. Thus,

$$d\,(\text{for } K = 0.3)$$

U ⟍ V	0	1/3	1/2	2/3	
0	1	1	1	1	1
1/3	1	0.994	0.988	0.992	1
1/2	1	0.997	1.000	0.998	1
2/3	1	1.015	1.009	1.012	1
1	1	1	1	1	1

Figure 70 Figure 68 recomputed using a variable interpolation factor.

Comparison of Figs. 68 and 70 reveal that the use of an appropriate K factor to influence interpolation generally gives a better result to the objective of developing a spherical shape from a set of boundary curves. The determination of the appropriate K can easily be accomplished at a graphic console by the simple expedient of varying K gradually and by viewing the contours of the resulting surface. The more conventional cubic blending as described by S. A. Coons[2] is a special case of this variable surface at $K = 1/2$. In other words, implementation of Eq. (109) in one or both directions of the interpolation increases the options over the basic surface-patch technique; it is less general than the Coons correction surface explained in the reference; it is much more manageable by most users than would be the correction surface; and it is easy to build-in an automatic default to the basic Eq. (107) if the user is not trained to profitably use the available flexibility. The controlled test using a spherical surface does not, of course, prove categorically that variable interpolation significantly enhances the matching of any conceived shape. However, the principles and limited evidence indicate the high probability of such results. The use of computers in general and graphics in particular offers many opportunities to explore this concept in more depth with little expenditure of resources.

[2]Coons, "Surfaces for Computer-Aided Design of Space Forms."

EXERCISES

1. Redraw Fig. 67 changing curve D such that it intercepts the Y axis as before but now bisects the XZ coordinate plane. This creates a 1/16 spherical segment. Determine the adjusted equations of X, Y, and Z versus V. Calculate X, Y, Z, and d for $U = V = 1/2$ (using cubic interpolation). How does this compare to d at $U = V = 1/2$ in the table of Fig. 68?

2. Using Eq. (109), can you find a K for interpolation in the V direction that will make d (of Exercise 1) for $U = V = 1/2$ closer to unity?

3. Referring to Section D of this chapter and Section G-4 of Chap. II, consider Fig. 58. Using the four boundary curves, set up parametric equations for X, Y, and Z versus U and V. Using the Coons surface of Eq. (107) and cubic interpolation of Eqs. (99) and (99a), select $U = V = 1/2$. Solve for Y. Then determine X and Z. Using the Y solution, derive the equation of the generalized ellipse. Using that equation, input the X of the Coons surface and solve for Z. How does the Z from the generalized ellipse compare to the Z from the Coons surface? Would you think the Coons surface technique might be a viable alternative to the basic interpolation of Section D?

4. Using the formulas of Exercise 3, find two UV coordinate pairs that make X equal 1.6. (Hint: Conventional interpolation may be used in a trial and error process.)

5. Construct four approximately smooth (small wiggles) space curves by specifying six coordinate sets for each curve. Use parametric techniques and fit a least squares cubic versus arc length for each variable for each curve. Arc length may be approximated by chord lengths from point to point. Use Eq. (107) and select several pairs of UV coordinates to get surface points according to conventional cubic interpolation. Introduce a K factor of interpolation, as in Eq. (109), for each direction. Assign K values (between zero and one) for each direction, use U and V pairs from the previous selected set, derive X, Y, and Z for each pair of K's, and compare them to those derived from the conventional interpolation. (This exercise is a bit lengthy for manual manipulation and, therefore, would be better performed via programming.) If programmed, plot the XYZ surface for increments of 0.2 in U and V both for conventional fitting and for selected pairs of K's.

G. SURFACE INTERSECTIONS

One of the most extensive requirements in design and analysis applications is the need to portray intersecting surfaces. Characteristically, one may wish to view the trace of a planar cut through a surface where the plane is either parallel to one of the principal coordinate planes or where the plane is oblique to the coordinate planes. When two or more general surfaces are involved, we often have the problem of visualizing clearances. The clearance visualization may be enhanced by viewing the trace of each involved surface in one or more planes. Thus, most of the practical problems may be reduced to the problem of deriving traces of the intersections of one or more specified planes with a particular surface. The surface of Fig. 58 and the contours of Fig. 60 exemplify the cutting of a surface by planes—planes parallel to the XZ plane in this example.

Where surface equations exist, the problem is normally rather straightforward; for example, a sphere of equation

$$X^2 + Y^2 + Z^2 = 1$$

is cut by a plane of equation $X = 1/2$. The resulting trace in the YZ plane is simply derived by substitution to be

$$Y^2 + Z^2 = 3/4,$$

which we recognize as a circular cross section. If the sphere is to be intersected by the plane $2X + Y + 3Z = 1$, then the substitution of X in terms of Y and Z into the equation of the sphere gives an equation in Y and Z which is an ellipse. You may say that any plane that passes through a sphere will produce a circle, not an ellipse. Indeed, that is correct within the plane. However, what we derive by the process of variable elimination through substitution is an equation of the *projection* of the 3-D space curve into a 2-D principal plane. Thus, the circle in space projects as an ellipse, in general. If we were interested in seeing not the projection of YZ but the specific trace of YZ at a particular X, we would substitute that X in both equations. Both the resulting YZ curves are in that X plane. It is, as we said earlier, equivalent to cutting two surfaces (the sphere and the given plane) by a third surface that has the special property of being a plane parallel to one of the principal coordinate planes. Thus, for a plane at $X = 1/2$, the spherical equation reduces to $Y^2 + Z^2 = 3/4$. The given plane reduces to $Y + 3Z = 0$. This yields the two solutions—$Y = 0.822$, $Z = 0.274$ and $Y = -0.822$, $Z = 0.274$. These are the only two points of the two surfaces that are in common at $X = 1/2$. We could ascertain clearances for Z versus Y at $X = 1/2$ by simply plotting the two YZ equations above. By taking a series of X values between $X = 0$ and $X = 1$ (the range of X for the sphere), we would derive a corresponding series of YZ solutions. Then we could plot any variable as a function of either of the other two variables which give 2-D projections. We could also plot X, Y, and Z for the set of solutions, in a typical orthographic 3-D representation. The important point here is that this numerical technique will apply to general surfaces irrespective of whether or not closed form equations (as in our example) exist. To be more specific, we again refer to Fig. 58 of Section D. There we developed a procedure to get Z versus X for any Y coordinate. Suppose we want to see what Z versus Y looks like for a specified X. We do not have a single equation that represents the entire surface. The numerical solution process would proceed as follows:

First, we select some X where we would like to derive a cross section. Of course, we must pick an X within its possible range, which is zero to two in the present case. We select $X = 1.5$ for our illustration. Next we must use increments of Y, solve for the generalized elliptic equation of Z versus X for each Y as described in Section D, input $X = 1.5$, and finally solve for Z. Thus, we will derive a Z coordinate for each increment of Y, specifically for $X = 1.5$.

We note from the linear equation for curve B that $Y = 1.5$ when $X = 1.5$. At that point on curve B, $Z = 0$. Figure 58 shows that for a Y less than 1.5, X cannot be 1.5. Conversely, for a Y greater than 1.5, the generalized ellipse goes from $X = 0$ to some value of X greater than 1.5. Thus, a solution exists for all Y values from 1.5 to 3.0. For this illustration, we select Y in 0.5 increments.

From cubic interpolation shown by Fig. 59, we determine the percentage of the diagonal (Z versus X) for $Y = 1.5$, 2.0, 2.5, and 3.0, which is 1/2, 2/3, 5/6, and 1 as a fraction of the total Y interval. This fraction when input to Eq. (99), multiplied by (0.9 − 0.707), and added to 0.707 gives the proper percentage diagonal. Thus,

Fraction of Y Interval	Percentage of Diagonal	XZ Intercepts
1/2	0.803	1.50
2/3	0.850	1.67
5/6	0.886	1.83
1	0.900	2.00

From Section G-4 on generalized ellipses and Eqs. (84), (84a), and

(84b), we derive the equations that correspond to the data in the table above:

$$X^{3.16} + Z^{3.16} = 3.60$$
$$X^{4.26} + Z^{4.26} = 8.89$$
$$X^{5.72} + Z^{5.72} = 31.71 \tag{110}$$
$$X^{7.03} + Z^{7.03} = 130.7$$

In these equations, we insert $X = 1.5$ and derive $Z = 0$, 1.31, 1.71, and 1.96, respectively for $Y = 1.5$, 2.0, 2.5, and 3.0. Following this procedure, we could get as many points for our surface slice as we desire within practical computation limits. In the same sense, we may derive as many slices as we wish.

Figure 60 shows a series of slices. If a sufficient number of such cuts are developed, we need only pick X along the abscissa of the figure and read Z's from the contours.

Now, suppose the surface of Fig. 58 were cut by an oblique plane instead of an orthogonal plane. Then for any Y coordinate, we would derive an equation of a line in terms of X and Z in the plane. For the same Y coordinate, we would develop the equation of Z versus X as we have just described. In our example, this would be the equation of a general ellipse. The line and the general ellipse may or may not intersect. If they do intersect, there may be more than one point of intersection in the general case (maximum of two for the example of Fig. 58). Thus, one or more points of penetration result, but a continuous curve will not exist for each Y unless the cutting plane is parallel to the XZ coordinate plane. For successive values of Y, such points are joined together to form a space curve that will lie in the cutting plane.

To illustrate this principle, again consider the design depicted by Fig. 58. We will cut that surface with a plane that passes through the three (X, Y, Z) points of $(-1, 0, 0)$, $(1, 1, 2)$, and $(2, 3, 0)$. (The reader may wish to sketch the three points on the figure for visualization.) From Section C-2 of this chapter and Eq. (90), the three points give the plane equation

$$2X - 2Y - Z = -2 \tag{111}$$

Equation (110) of the previous example gives the four equations in X and Z that correspond to $Y = 1.5$, $Y = 2$, $Y = 2.5$, and $Y = 3$, respectively. From Eq. (111), we substitute those values of Y and derive the linear traces in X and Z:

$$\text{For } Y = 1.5, \quad 2X - Z = 1$$
$$\text{For } Y = 2.0, \quad 2X - Z = 2$$
$$\text{For } Y = 2.5, \quad 2X - Z = 3 \qquad (112)$$
$$\text{For } Y = 3.0, \quad 2X - Z = 4$$

Equations (110) and (112) are paired to derive the XZ intersections for the successive Y values. Thus $X^{3.16} + Z^{3.16} = 3.60$ is paired with $2X - Z = 1$. The solution, by iteration, is approximately

$$X = 1.13, \quad Z = 1.27$$

The next pair is

$$X^{4.26} + Z^{4.26} = 8.09$$

and

$$2X - Z = 2$$

The solution is approximately

$$X = 1.58, \quad Z = 1.16$$

The third pair is

$$X^{5.72} + Z^{5.72} = 31.71$$

and

$$2X - Z = 3$$

The solution is

$$X = 1.83, \quad Z = 0.66$$

(Because of the steep slope of the elliptic curve, the X solution is almost the same as the X intercept.) Similarly, for $Y = 3.0$, the solution is $X = 2$ and $Z = 0$ (which is one of the given planar points).

To derive some additional coordinates, we set $Y = 3/4$ and derive $X = 0.36$ and $Z = 1.22$. It is noted that for $Y = 0$, the planar trace is $2X - Z = -2$, which means that when $X = 0$, $Z = 2$, and when X increases, Z increases. Reference to Fig. 60 shows that this trace will not intersect the circle. Without going into more detail, we can determine from boundary conditions that the first solution for increasing Y occurs when $Y = 3/8$. This gives the solution

$$X = 0, \quad Z = 1.125$$

which is an end point of the curve formed by intersecting the surface of Fig. 58 with the oblique illustrative plane of Eq. (111).

We summarize the resulting coordinates for the space curve of intersection as follows:

Y	X	Z
0.375	0	1.125
0.75	0.36	1.22
1.50	1.13	1.27
2.00	1.55	1.15
2.50	1.83	0.66
3.00	2.00	0

Of course, we are at liberty to plot the projections of this curve in any 2-D coordinate plane we wish, or we may wish to make a plot in a 3-D orthographic projection.

In the first case, we will plot the projection of Z versus X as in Fig. 71. Then we make a 3-D plot of the same curve, which is shown in Fig. 72.

In the previous section, we described certain procedures to develop surfaces from four space curves. We found it convenient to generate such surfaces through the medium of parametrics—letting each of the variables be functions of both U and V. Now we will prescribe a process by which such surfaces may be cut with planes to produce a curve of intersection. In particular, we will select planes parallel to the principal coordinate planes; that is, $X = a$, $Y = b$, or $Z = c$, where a, b, and c are constants. It is important to

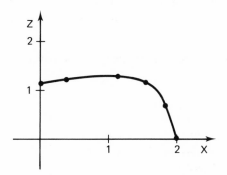

Figure 71 Projection of space curve formed by cutting the surface of Fig. 58 with the oblique plane, $2X - 2Y - Z = -2$.

understand that picking U or V to be constant would not, in general, produce curves in the principal planes. Reference to Fig. 65 reminds us that selections of increments of U and V produce space curves in X, Y, and Z. Figure 65 shows three views of the projections of such curves.

We will examine a portion of the tabular data from which Fig. 65 is constituted. The numbers in the body of the tables are not the actual fairing data that the figure represents, but they are proportional.

X

V \ U	0	0.2	0.4	0.6	0.8	1.0
0	19.8	14.85	10.13	5.71	1.96	0
0.2	19.8	15.35	11.41	7.84	4.79	3.13
0.4	19.8	16.04	12.49	9.88	7.32	5.88
0.6	19.8	16.75	14.06	11.61	9.39	8.10
0.8	19.8	17.30	14.97	12.77	10.71	9.50
1.0	19.8	17.52	15.32	13.20	11.19	10.00

Y

V \ U	0	0.2	0.4	0.6	0.8	1.0
0	5.00	4.96	4.75	4.25	2.88	0
0.2	5.28	5.21	4.95	4.41	3.11	0.50
0.4	5.52	5.46	5.22	4.79	3.81	1.90
0.6	6.62	6.53	6.27	5.88	5.21	4.12
0.8	8.16	8.08	7.88	7.62	7.29	6.92
1.0	10.00	10.00	10.00	10.00	10.00	10.00

Similarly, we prescribe the range of Z's without filling the table, thus:

Z

V \ U	0	→	1
0	6.2	→	0
↓	↓		↓
1	11.3	→	11.3

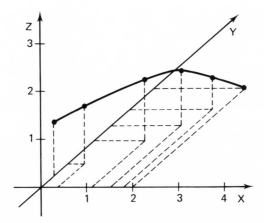

Figure 72 Plot of 3-D space curve of the surface of Fig. 58 cut by the plane $2X - 2Y - Z = -2$.

Upon careful inspection, we may observe that the first and last columns and the first and last rows within each table are the X, the Y, and the Z coordinates of the surface boundary curves—denoted as A, B, C, and D in Fig. 65. (This is true, although the pattern of columns and rows in the table do not conform to the pattern on the figure.) To describe this, note that the first column of the X's is constant at 19.8. Look at the plan view of Fig. 65 and regard the upper left corner at the intersection of curves B and C as the origin for the X's. Thus, the first column is curve D. Starting at the corner of curve D, we note that the first row of the X's goes from 19.8 to zero—that is curve C on the figure. The last row of the table (which is connected to the other end of the column for curve D) varies from 19.8 to 10, which is curve A. The last column varies between zero and ten, which is curve B. Note that the contours connect curve D to B and curve A to C, which is the way the interpolation flows within the table.

Suppose we want to use the data of Fig. 65 (from the tables just presented) to see what the fairing surface looks like at one or more stations (X locations). For each X of interest, the procedure would be to use the UV table of X's and find, through inverse interpolation, a set of UV coordinates that produces the desired X value.

For each (U, V) of the set, the other tables are used with direct interpolation to find both Y and Z. Thus, a set of YZ coordinates is evolved for each X location. Let us see how the first step (the inverse interpolation) proceeds. We could take increments of one of the parameters and vary the other for each increment until the specified X (or Y or Z) is determined. Suppose we want to find Z versus Y for $X = 18$ of the tabular data. If we vary V for increments of $U = 0.2$, there will be no column where X crosses 18. Alternatively, if we use increments of V and vary U, we find that each row yields 18. It happens to be between $U = 0$ and $U = 0.2$ for each V. Regular interpolation would give a good approximation of the value of U in each case.

   Due to the nature of surfaces, we can never be sure of which way to carry out this numerical process. If we want to derive U and V for $X = 8$ for instance, there are three rows (U variation) and three columns (V variation) that will cross the value $X = 8$ and therefore, can yield a total of six UV pairs to give $X = 8$ by interpolating within rows and within columns. Because of this uncertainty, we will ensure the maximum number of UV values for a given X by going both ways through the table; that is, using increments of U and varying V in the table and using increments of V and varying U. Using increments of 0.2 systematically as prescribed, we would produce 36 UV pairs to consider. We must go through the entire table, in general, because there could be more than one crossing on a row or column. If this process does not produce enough UV pairs, we may have to reconstitute each table

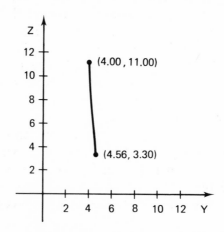

Figure 73   Cross section plot of Fig. 65 for $X = 8$ (one of the contours in Fig. 74).

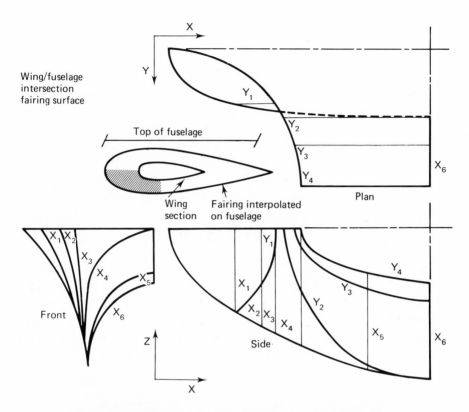

Figure 74 Same as Fig. 65 except for contours of constant cross section.

with a finer grid, for example, increments of 0.1 or 0.05. The number of grid units can be tied to the number of *UV* pairs that are derived by appropriate programming; for example, we may wish to increase the number of increments until we derive at least five *UV* pairs. Each *UV* pair is used, as previously described, to generate a corresponding *Y* and *Z*. If we use $X = 8$ to illustrate the process and use cubic interpolation, we derive from the table of *X*'s the following set of *UV* coordinates:

$$(0.495, 0), \quad (0.59, 0.2), \quad (0.74, 0.4)$$
$$(0.6, 0.215), \quad (0.8, 0.46), \quad (1.0, 0.59)$$

Study the table of *X*'s to verify the reasonableness of these *UV* pairs for $X = 8$. The corresponding *YZ* coordinates from the other tables are

$$(4.56, 3.30), \quad (4.44, 5.29), \quad (4.18, 7.64)$$

$$(4.42, 5.65), \quad (4.14, 8.17), \quad (4.00, 11.0)$$

Some of the points are close to others because of the proximity of the solutions to the arbitrary grid.

The plot of Z versus Y for $X = 8$ is shown in Fig. 73.

For other values of X, a stacking of cross sections can similarly be derived. Figure 74 is the same as Fig. 65 except that YZ contours for specific values of X and XZ contours for specific values of Y are depicted. Figure 73 depicts one of the set of contours shown as the front view of Fig. 74.

Thus, four space curves can be used to produce a parametric surface with space contours as in Fig. 65 or with cross section contours (in a principal plane) as in Fig. 74. With graphics and with variable interpolation (explained in Section F), considerable insight may be obtained as surfaces are *varied* to fit a *fixed* set of boundaries. The cross section cuts are quite useful in further design and/or analysis. Typical requirements are for arc lengths, areas, and volumes, which will be treated in the next section.

H. LENGTHS OF CURVES, AREAS, AND VOLUMES

As we have said in many ways throughout the text, a curve is either constituted from a set of points representing numerical data or from describing equations from which points may be derived. In either case, a sequence of points are successively joined by straight lines to make a CRT display or a hard-copy plot. The density of points will be determined by the many considerations that we have described. The emphasis in this section is on the short lines (chords) that connect the points and their implication. Ideally, we may desire to learn the exact length or perimeter of a curve and the area within its confines. With such data, a series of surface cuts, producing a corresponding series of perimeters and areas, may be integrated over an interval of interest to derive surface area and volume, respectively. We will not discuss the process of integration here because that is well treated in texts on numerical analysis, and because the alternative curve-fitting techniques, which underlie numerical processes, were described in Chap. II. The numerical process of integration will induce some error depending on the density or frequency of data points along the curve and on the degree of accurate representation by curve fitting. The question we will discuss here is, to what extent does the density or frequency of points affect the curve length and the area computation at a single location, that is, in 2-D?

Suppose a surface of an object is cut by a plane. The intersection will be a closed curve that circumnavigates the surface such as is shown in Fig. 75.

We will discuss one quadrant of the complete curve, although the principles are, of course, extendable to the entire curve. The length of the curve from A to B of Fig. 75 is approximated by summing the lengths of the chords. Chords are always equal to (linear curves) or less than the lengths of the arcs they approximate. Thus, their sum falls short of the true length. This is true irrespective of whether or not inflections exist on the curve. It is, therefore, evident that the number of points and their chord lengths affect the accuracy of determining curve lengths, which are, in turn, used to derive surface areas. More will be said about the degree of accuracy for curve lengths shortly. (It is important to realize that any study of "accuracies" is independent of aesthetic or styling requirements.)

The area within a 2-D closed curve can be determined by summing determinants; however, the form is so simple that we will give the expanded algebraic expression for the area inside a closed polygon. Suppose the closed curve is approximated by N points—(X_1, Y_1), (X_2, Y_2), . . . , (X_N, Y_N), where the first point is

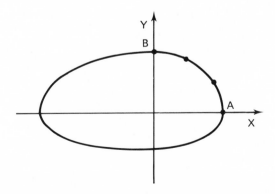

Figure 75 Example of surface cut by a plane to give a 2-D closed curve.

arbitrary (start anywhere) and the sequence is counterclockwise. Then,

$$\text{Area} = \frac{1}{2}\left[(X_1Y_2 + X_2Y_3 + \cdots + X_{N-1}Y_N + X_NY_1)\right.$$

$$\left. - (X_2Y_1 + X_3Y_2 + \cdots + X_NY_{N-1} + X_1Y_N)\right] \tag{113}$$

This formula is more general than standard formulas for integration since it applies to complete closed curves. Notice that the first point is repeated. The polygon of N points will always lie entirely within the curve if the curve always bends toward the origin (assuming the origin is inside the curve). On the other hand, suppose the curve is composed of four quadrants, each of which is a generalized ellipse (Chap. II, Section G-4) with an exponent less than unity. In such a case, the polygon will lie entirely outside the curve. For curves with inflection points, both conditions exist; that is, some chords are inside the curve and some are outside, and this produces a degree of compensating errors in area determination. We will assume the worst case, from an approximation point of view, where the error is all in one direction. This is the case in Fig. 75. For Fig. 75, we could take N points around the entire periphery and apply Eq. (113). If we are only interested in the first quadrant area, we could take $N - 1$ points between A and B plus $(0, 0)$ as the Nth point.

Now let us turn to the problem of determining how many points give what kind of accuracy for both arc length and area. Suppose, for example, we consider a circle with a unit radius and a center at the origin. Consider the first quadrant and that the 90° arc is approximated by two chords, $(1, 0)$ to $(0.707, 0.707)$ and $(0.707, 0.707)$ to $(0, 1)$. The sum of their lengths is 1.531 as compared to the arc length of 1.571. This means that with the three points, the chord-length sum is within 2.4% of the arc it approximates. For the area in the first quadrant, we add the point $(0, 0)$ and apply Eq. (113); thus,

Equation (113) is the continued 2-D determinant above, where the products along the right pointing arrows are added and those to the left are subtracted. If this is not clear, apply Eq. (113) directly using

(0, 0) as one of the points. This area of 0.707 falls short of the area of a quarter of a circle, which is 0.785 or 11.1% greater. We see from this simple example that the error from a given set of points is somewhat greater when approximating area than it is when approximating lengths. This is always the case when there are no inflections on the curve. We feel that it would be instructive to present the relationship between actual lengths and areas in 2-D in comparison to those derived from chords for N points. We have done some calculations of these ratios for certain elementary and known mathematical forms. The following table results:

1. For circles with equi-spacing of points via increments of θ, (1st quadrant):

Number of Chords, N	Arc Length S	Sum of Chords C	S/C %	Ratio, True Area to Chord Area, %
1	1.571	1.414	11.0	
2		1.532	2.4	
3		1.548	1.8	
4		1.560	0.7	1.2

2. For $Y = \sin X$ from 0 to $\pi/2$ with equi-spacing on X:

1	1.910	1.862	2.6	
2		1.895	0.8	5.5

3. For parabola $Y = 1 - X^2$, equi-spacing on X from 0 to 1:

1	1.479	1.414	4.6	
2		1.460	1.3	
3		1.470	0.6	2.9

4. For the first quadrant of a circle with equi-spacing of X for $N = 4$,

$$\frac{S}{C} = 1.2\%$$

and

$$\text{area ratio} = 5.5\%$$

Certainly equi-spacing of points, if easily obtained, results in a requirement for fewer points to achieve a certain accuracy as seen in comparing cases (1) and (4). It is also apparent that area computation is more sensitive to N than arc length is. However, of the few cases treated here for illustration, the maximum error is slightly over 5% for areas when $N = 4$. Since we were working with only one quadrant, we can extrapolate and say that 16 chords (or 17 points) would give "fair" accuracy for the entire perimeter and area computation in 2-D. That is, of course, true to the extent that the functions selected are representative in terms of a sensitivity analysis. We believe that they are. Obviously the nature of the application should dictate the level of accuracy and hence the size of N. The preceding tables imply that, in many practical cases, one need not be concerned with equi-spacing of points versus equi-spacing of just the abscissa, X. Also, it should be clear that, except perhaps for certain aesthetic needs and needs for certain computations that use the approximating points, vast numbers of points are not required to get good approximations of arc length and area which are, in turn, used to derive surface areas and volumes. The implementation of these techniques is particularly useful in interactive graphic design and analysis. They have been successfully used to automate derivation of areas and volumes via graphic design for airplane conceptual design.

EXERCISES

1. Consider the ellipse $X^2/9 + Y^2/4 = 1$. In the first quadrant, plot nine points (eight chords) with equi-spacing on the X axis. Find the total length of the eight chords as the *standard* representation. Now use four chords and compute the ratio of S/C to determine the accuracy of the approximation to the standard. Use the nine points and the origin or calculus integration to get the area under the curve. Use the four-chord approximation and the corresponding points to find the approximate area. How close does it approximate the standard?

2. Use the ellipse of Exercise 1 at $Z = 0$, $X^2/16 + Y^2/36 = 1$ at $Z = 2$, and $X^2/64 + Y^2/144 = 1$ at $Z = 4$.
   Use four chords per quadrant to approximate total elliptic lengths and total area at each Z. From the three sets of approximations, make a graph for both length and area versus Z. Integrate numerically and derive total surface area and volume between $Z =$ zero and $Z = 4$.

APPENDICES

Appendix A:
DERIVATION OF FORMULAS TO DEFINE
A CIRCLE OF GIVEN RADIUS,
TANGENT TO TWO OTHER CIRCLES.

Suppose a given circle equation has had its center decreased by h and k such that it now has the form

$$X^2 + Y^2 = r_1^2 \qquad \text{(A-1)}$$

Moving the circle does not affect the radius. Translating the second circle also by (h, k) gives a new center, which we shall denote as (a, b). Therefore, the relative positions of the two circles, wherever they were originally, will be unchanged. The equation of the second circle is

$$(X - a)^2 + (Y - b)^2 = r_2^2 \qquad \text{(A-2)}$$

We use Eqs. (A-1) and (A-2) to derive the center of the desired circle and then increase its coordinates by (h, k) to account for the original translation. This in no way changes the solution technique and the ultimate result, but it does simplify the formulation.

Suppose we specify r_3 to be the radius of the circle to be derived. Since this circle must be tangent to the given circles, its center must be $r_1 + r_3$ from the center of the first and $r_2 + r_3$ from the center of the second. In other words, if we increase the radius of Eqs. (A-1) and A-2), we will have two intersecting circles (if a solution exists) where the center to be derived (X_C, Y_C) is at one of the two intersections. So, we rewrite Eqs. (A-1) and (A-2) with the expansion of Eq. (A-2). Thus, we have

$$X^2 + Y^2 = (r_1 + r_3)^2 \qquad \text{(A-3)}$$

$$X^2 - 2aX + a^2 + Y^2 - 2bY + b^2 = (r_2 + r_3)^2 \qquad \text{(A-4)}$$

Subtracting Eq. (A-4) from (A-3) gives

$$2aX + 2bY - a^2 - b^2 = (r_1 + r_3)^2 - (r_2 + r_3)^2$$

or

$$X = \frac{(r_1 - r_2)(r_1 + r_2 + 2r_3) + a^2 + b^2}{2a} - \frac{b}{a} Y$$

We define

$$K_1 = \frac{(r_1 - r_2)(r_1 + r_2 + 2r_3) + a^2 + b^2}{2a}$$

and

$$K_2 = -\frac{b}{a}$$

Thus, we have

$$X = K_1 + K_2 Y \tag{A-5}$$

This is the common chord that intersects the circles to give the solution.

Equation (A-5) is substituted in (A-3) to give,

$$(K_1 + K_2 Y)^2 + Y^2 = (r_1 + r_3)^2$$

or

$$(K_2^2 + 1) Y^2 + 2K_1 K_2 Y + K_1^2 - (r_1 + r_3)^2 = 0$$

$$Y^2 + \frac{2K_1 K_2 Y}{K_2^2 + 1} + \frac{K_1^2 - (r_1 + r_3)^2}{K_2^2 + 1} = 0 \tag{A-6}$$

We define

$$K_3 = \frac{2K_1 K_2}{K_2^2 + 1}$$

and

$$K_4 = \frac{K_1^2 - (r_1 + r_3)^2}{K_2^2 + 1}$$

Then Eq. (A-6) becomes

$$Y^2 + K_3 Y + K_4 = 0$$

The quadratic formula is applied to give

$$Y = \frac{-K_3 \pm (K_3^2 - 4K_4)^{1/2}}{2}$$

These two solutions are defined as K_5 and K_6; so the corresponding X solution referring back to Eq. (A-5) is

$$X = K_1 + K_2 K_5$$

or

$$X = K_1 + K_2 K_6$$

depending on which value of Y is selected according to selection criteria described in the text. Finally, these solutions for X and Y are incremented by h and k, respectively, to locate (X_C, Y_C) of the desired arc. Its equation becomes

$$(X - X_C)^2 + (Y - Y_C)^2 = r_3^2$$

Appendix B:
FINDING THE EQUATION OF A CIRCLE
INSCRIBED IN A TRIANGLE

The sides of a triangle may be defined either by three vertices from which the equations of the sides are derived or from explicit equations. The center of an inscribed circle must lie on the bisectors of each of the interior angles of the triangle. The center location is derived by finding the intersection of any two of the three bisectors. The three equations of the sides are in the form

$$A_1X + B_1Y + C_1 = 0$$
$$A_2X + B_2Y + C_2 = 0$$
$$A_3X + B_3Y + C_3 = 0$$

The equation of a bisecting line is determined by equating the distances from the angle's boundary lines to a variable point (X, Y), (the general point on the bisecting line). The distance formula, from analytic geometry, is

$$d = \pm \frac{AX + BY + C}{\pm(A^2 + B^2)^{1/2}}$$

where the sign of the denominator is the same as that of the B coefficient (or opposite to the sign of the slope), and the sign in front of the fraction depends on whether the point is above or below the line (as measured from line to point). If we arbitrarily take the first and second triangle equations and equate the distances, we have

$$\frac{A_1X_1 + B_1Y_1 + C_1}{(A_1^2 + B_1^2)^{1/2}} = \pm \frac{A_2X + B_2Y + C_2}{(A_2^2 + B_2^2)^{1/2}}$$

We define

$$K_1 = (A_1^2 + B_1^2)^{1/2}$$

and

$$K_2 = (A_2^2 + B_2^2)^{1/2}$$

Thus, we have either

$$(A_1K_2 + A_2K_1) + (B_1K_2 + B_2K_1) + (C_1K_2 + C_2K_1) = 0 \quad \text{(B-1)}$$

or

$$(A_1K_2 - A_2K_1) + (B_1K_2 - B_2K_1) + (C_1K_2 - C_2K_1) = 0 \quad \text{(B-2)}$$

The decision as to which of the two equations applies depends on the signs of the slopes of the two given lines and the signs of the two distances. If an odd number of signs (of the four slope and distance signs) is negative, then Eq. (B-1) applies. Otherwise, Eq. (B-2) applies.

We determine the algebraic sign of the distance by using the vertex opposite to the line being used. Thus, if we are using $A_1X + B_1Y + C_1$, then we use (X_3, Y_3). This is done by determining Y from the equation at X_3 and noting whether Y_3 is less than or greater than that value. This works because the vertex lies in the same direction as any point on the bisector. To illustrate this distance-sign determining process, suppose the line has equation $2X - Y + 4 = 0$, and the opposite vertex is $(2, 3)$. Then for $X = 2$, $Y = 8$; so the Y coordinate of the point is less than the corresponding Y of the line. Hence, the distance from the line to the point is a negative distance. This process is followed when deriving the "sign" of the distance from each line used. If the other two equations were $X + 2Y - 8 = 0$ and $3X - Y - 3 = 0$, and if we want to first get the equation of the bisector of the angle formed by $2X - Y + 4 = 0$ and $X + 2Y - 8 = 0$, we have to decide on Eq. (B-1) or (B-2). The two slopes have one positive and one negative sign. In this case, the two distances from the lines to opposite vertices also have one positive and one negative sign. Therefore, according to the rule of signs just presented, formula (B-2) would apply in deriving one bisector equation. A similar process at one of the other angles leads to another bisector equation. Solving the two simultaneously gives the circle center, (X_C, Y_C). Then, any of the lines may be used to get the radius; for example,

$$r^2 = \frac{(A_1X_C + B_1Y_C + C_1)^2}{A_1^2 + B_1^2} \tag{B-3}$$

Then the circle equation becomes

$$(X - X_C)^2 + (Y - Y_C)^2 = r^2 \tag{B-4}$$

Appendix C:
FINDING THE EQUATION OF A CIRCLE
THAT IS TANGENT TO TWO OTHER CIRCLES
AND PASSES THROUGH A GIVEN POINT

This case is depicted in Fig. 23 of the text and characteristically has two mathematical solutions for practical input data. A CRT display lends itself to this kind of input. We will define the two given circles such that their centers have coordinates (x_1, y_1) and (x_2, y_2) with radii of r_1 and r_2. The given point has coordinates (X_3, Y_3). We first make a simple translation by X_3 and Y_3, which has no effect on the relative geometry but simplifies the solution process and the resulting formulas. After the translation, the centers of the two circles are (X_1, Y_1) and (X_2, Y_2) and the given point is at $(0, 0)$.

After we have solved for the coordinates of the center of the desired circle and its radius, we will make the simple inverse translation. The center of the desired circle can be described by the following three relationships after translation in terms of the unknown radius R. Inspection of Fig. 23 may assist in understanding them. The logic is similar to that in the text for case (b) in Chap. II, Section G-2. Thus,

$$(X - X_1)^2 + (Y - Y_1)^2 = (r_1 + R)^2 \qquad \text{(C-1)}$$

$$(X - X_2)^2 + (Y - Y_2)^2 = (r_2 + R)^2 \qquad \text{(C-2)}$$

$$X^2 + Y^2 = R^2 \qquad \text{(C-3)}$$

First, we use R in terms of X and Y from Eq. (C-3) and substitute into Eqs. (C-1) and (C-2). This gives

$$(X - X_1)^2 + (Y - Y_1)^2 = \left(r_1 + \sqrt{X^2 + Y^2}\right)^2$$

and

$$(X - X_2)^2 + (Y - Y_2)^2 = \left(r_2 + \sqrt{X^2 + Y^2}\right)^2$$

Expanding and simplifying, we have

$$-2X_1 X - 2Y_1 Y + X_1^2 + Y_1^2 = r_1^2 + 2r_1\sqrt{X^2 + Y^2} \qquad \text{(C-4)}$$

and

$$-2X_2 X - 2Y_2 Y + X_2^2 + Y_2^2 = r_2^2 + 2r_2\sqrt{X^2 + Y^2} \qquad \text{(C-5)}$$

We remove the radical by multiplying the first of these equations by r_2, the second by r_1, and then by subtracting the second from the first. This gives

$$(2X_2r_1 - 2X_1r_2)X + (2Y_2r_1 - 2Y_1r_2)Y$$
$$+ r_2(X_1^2 + Y_1^2 - r_1^2) - r_1(X_2^2 + Y_2^2 - r_2^2) = 0 \quad (C\text{-}6)$$

Now in Eq. (C-6), we define

$$K_1 = 2X_2r_1 - 2X_1r_2$$
$$K_2 = 2Y_2r_1 - 2Y_1r_2$$
$$K_3 = r_2(X_1^2 + Y_1^2 - r_1^2) - r_1(X_2^2 + Y_2^2 - r_2^2)$$

So Eq. (C-6) is simplified to

$$K_1X + K_2Y + K_3 = 0 \quad (C\text{-}7)$$

Now we solve for X in terms of Y from Eq. (C-7) to get

$$X = -\frac{K_2}{K_1} Y - \frac{K_3}{K_1}$$

We define

$$K_4 = -\frac{K_2}{K_1}$$

$$K_5 = -\frac{K_3}{K_1}$$

So,

$$X = K_4Y + K_5 \quad (C\text{-}8)$$

Now we use Eq. (C-8) in either Eq. (C-4) or (C-5); (we pick Eq. (C-4) arbitrarily). This gives

$$- 2X_1(K_4Y + K_5) - 2Y_1Y + X_1^2 + Y_1^2$$
$$= r_1^2 + 2r_1\sqrt{(K_4Y + K_5)^2 + Y^2}$$

or

$$\frac{(-X_1K_4 - Y_1)Y}{r_1} + \frac{(-2X_1K_5 + X_1^2 + Y_1^2 - r_1^2)}{2r_1}$$
$$= \sqrt{(K_4^2 + 1)Y^2 + 2K_4K_5Y + K_5^2} \quad (C\text{-}9)$$

Now we define

$$K_6 = \frac{-X_1 K_4 - Y_1}{r_1}$$

and

$$K_7 = \frac{-2X_1 K_5 + X_1^2 + Y_1^2 - r_1^2}{2r_1}$$

Then Eq. (C-9) may be rewritten as

$$K_6 Y + K_7 = \sqrt{(K_4^2 + 1) Y^2 + 2K_4 K_5 Y + K_5^2}$$

Squaring and collecting terms, we have

$$(K_6^2 - K_4^2 - 1) Y^2 + (2K_6 K_7 - 2K_4 K_5) Y + (K_7^2 - K_5^2) = 0 \quad \text{(C-10)}$$

Let

$$K_8 = K_6^2 - K_4^2 - 1$$
$$K_9 = 2K_6 K_7 - 2K_4 K_5$$
$$K_{10} = K_7^2 - K_5^2$$

Then Eq. (C-10) becomes

$$K_8 Y^2 + K_9 Y + K_{10} = 0 \quad \text{(C-11)}$$

We solve for Y by quadratic formula, as given in formula (45) of the text.

For each Y, we use Eq. (C-8) to get corresponding X coordinates for the center. Equation (C-3) is used to derive R. We simply add X_3 and Y_3 to this center to compensate for initial translation. The appropriate solution of the two is determined from the location of the initial tracking symbol on the display near the first circle, the terminal position of the symbol, and our clockwise convention needed to draw the desired arc. The convention process and successive computation of points is most easily conducted by translating the circle center to the origin, and the initial and terminal points will have corresponding translations. Then we establish clockwise rotation by the fact that (when the center is at the origin) for increasing X, the derived Y from the equation *must* be positive, and for decreasing X, Y coordinates *must* be negative (starting at the initial X and moving toward the terminal X). At the beginning, there are two candidate solutions; but only one will give the desired clockwise convention. Therefore, the solution selection can be established on the first solution for the Y's for the first X after the initial point. Some software systems use a counterclockwise convention. In any case, when successive coordinates have been developed for the circle centered at the origin, it is then merely necessary to retranslate each point by (X_C, Y_C) to establish the intended arc. Of course, other procedures could be derived in an analogous way to those presented here should another convention be preferred.

Appendix D:
DERIVATION OF INPUT SLOPE LIMIT
TO DERIVE AN ELLIPTIC EQUATION
FOR AN ELLIPSE IN STANDARD POSITION

For the case where an ellipse has its center on the Y-axis with horizontal and vertical axes, the basic equation is

$$\frac{X^2}{a^2} + \frac{(Y - k)^2}{b^2} = 1 \qquad \text{(D-1)}$$

We desire the equation to produce an arc from the horizontal orientation at the top to some location at the side as depicted below:

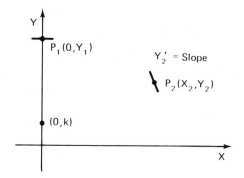

The horizontal slope eliminates what would have been a fourth unknown parameter, h. In the present case, we must solve for a, b, and k, the semimajor axis, the semiminor axis, and the location of the center, respectively. Therefore, we must input three constraints to set up three equations in three unknowns. The coordinates of P_1 and P_2 provide two conditions. The third is an input slope at P_2—a reasonable design requirement. We must determine the range of acceptable slopes for input. We define the slope to be input as Y_2'. If Y_2' were ∞ (vertical slope), we would have a full elliptic quadrant from P_1 to P_2. This is the trivial case. The parameter a would simply be X_2, the parameter k would be Y_2, and b would be $Y_1 - k$. If, on our diagram, Y_2' were positive, we would have more than a quadrant (two values of Y for some range of X's). We will not concern ourselves with this case. Therefore, we restrict Y_2' to be less than vertical and proceed to find the lower limit as follows. The equation of the derivative, derived from Eq. (D-1), is

$$Y' = - \frac{b^2 X}{a^2(Y - k)} \tag{D-2}$$

From Eq. (D-1) and the input points, we have

$$(Y_1 - k)^2 = b^2 \tag{D-3}$$

$$\frac{x_2^2}{a^2} + \frac{(Y_2 - k)^2}{b^2} = 1 \tag{D-4}$$

From Eq. (D-2),

$$Y_2' = \frac{- b^2 X_2}{a^2(Y_2 - k)} \tag{D-5}$$

From Eq. (D-4),

$$\frac{X_2^2}{a^2} = \frac{b^2 - (Y_2 - k)^2}{b^2}$$

Now substitute b^2 in terms of Y from Eq. (D-3). Thus,

$$\frac{X_2^2}{a^2} = \frac{(Y_1 - k)^2 - (Y_2 - k)^2}{b^2}$$

Factoring and multiplying by a^2 gives

$$X_2^2 = \frac{a^2}{b^2} (Y_1 - Y_2)(Y_1 + Y_2 - 2k)$$

or

$$\frac{b^2}{a^2} = \frac{(Y_1 - Y_2)(Y_1 + Y_2 - 2k)}{X_2^2} \tag{D-6}$$

Now substitute the expression for b^2/a^2 of Eq. (D-6) into Eq. (D-5). After simplification, we get

$$Y_2' = - \frac{(Y_1 - Y_2)(Y_1 + Y_2 - 2k)}{X_2^2} \cdot \frac{X_2}{Y_2 - k}$$

or

$$Y_2' = \frac{Y_2 - Y_1}{X_2} \cdot \frac{Y_2 + Y_1 - 2k}{Y_2 - k} \tag{D-7}$$

Since Eq. (D-7) is the relationship between the input Y_2' and the ellipse center k, the limiting value or boundary condition for Y_2' is derived by taking the limit of Y_2' as k approaches negative infinity. In other words, Eq. (D-7) shows us what happens to the input slope for continually decreasing locations of the center. Thus,

$$\lim_{k \to -\infty} Y_2' = \frac{2(Y_2 - Y_1)}{X_2}$$

or simply twice the slope of the line between P_1 and P_2.

Therefore, we can derive an elliptic equation with the stated restrictions if and only if the input slope is in excess of twice the slope (in magnitude) of the line between P_1 and P_2 and less than vertical. For example, suppose $P_1 = (0, 6)$ and $P_2 = (5, 5)$. Since the slope between P_1 and P_2 is $-1/5$, we can derive an elliptic equation for the desired arc as long as we input a slope at P_2 that is greater in magnitude than $2/5$. It cannot equal $-2/5$ since $-2/5$ is the limit.

Appendix E:
DERIVATION OF THE EQUATION OF
AN ELLIPSE IN STANDARD POSITION
WITH CENTER ON THE Y AXIS

This case is described by

$$\frac{X^2}{C_1} + \frac{(Y - k)^2}{C_2} = 1 \qquad \text{(E-1)}$$

The same form is in Appendix D except that C_1 and C_2 are used for a^2 and b^2, respectively. One point on the arc is $P_1(X_1, Y_1)$, which we select to be on the Y axis. The other point is $P_2(X_2, Y_2)$, and the given slope at P_2 is m. This is depicted in Fig. 30 in the text. The equation of the derivative, derived from Eq. (E-1) above is

$$Y' = -\frac{C_2 X}{C_1(Y - k)} \qquad \text{(E-2)}$$

Substituting the two points into Eq. (E-1) gives

$$\frac{0^2}{C_1} + \frac{(Y_1 - k)^2}{C_2} = 1 \qquad \text{(E-3)}$$

and

$$\frac{X_2^2}{C_1} + \frac{(Y_2 - k)^2}{C_2} = 1 \qquad \text{(E-4)}$$

Solving Eq. (E-2) for C_2 gives

$$m = -\frac{C_2 X_2}{C_1(Y_2 - k)}$$

$$C_2 = \frac{C_1 m(k - Y_2)}{X_2} \qquad \text{(E-5)}$$

Now substitute Eq. (E-5) into (E-3) and (E-4) to give

$$\frac{X_2(Y_1 - k)^2}{C_1 m(k - Y_2)} = 1 \qquad \text{(E-6)}$$

and

$$\frac{X_2^2}{C_1} + \frac{X_2(Y_2 - k)^2}{C_1 m(k - Y_2)} = 1 \qquad \text{(E-7)}$$

Now set Eq. (E-6) equal to (E-7) since the left sides of the two equations equal unity. Thus,

$$\frac{X_2(Y_1 - k)^2}{C_1 m(k - Y_2)} = \frac{X_2^2}{C_1} + \frac{X_2(Y_2 - k)^2}{C_1 m(k - Y_2)}$$

C_1, which is one of the three unknowns, drops out of the equation and we proceed to solve for k. So,

$$\frac{X_2(Y_1 - k)^2}{m(k - Y_2)} = X_2^2 + \frac{X_2(Y_2 - k)^2}{m(k - Y_2)}$$

or

$$(Y_1 - k)^2 = mX_2(k - Y_2) + (Y_2 - k)^2$$

Collecting terms gives

$$k(2Y_2 - 2Y_1 - mX_2) = Y_2^2 - Y_1^2 - mX_2Y_2$$

Thus,

$$k = \frac{Y_2^2 - Y_1^2 - mX_2Y_2}{2Y_2 - 2Y_1 - mX_2} \tag{E-8}$$

Equation (E-8) is used to derive k. Using this value of k, we return to Eq. (E-6) to solve for C_1. Thus,

$$C_1 = \frac{X_2(Y_1 - k)^2}{m(k - Y_2)}$$

Then using k from Eq. (E-8) and C_1 from Eq. (E-6), we use Eq. (E-5) to get C_2. Thus,

$$C_2 = \frac{C_1 m(k - Y_2)}{X_2}$$

Then C_1, C_2, and k are used in Eq. (E-4) to derive the final elliptic equation. An illustration of the use of these formulae is given in the text.

Appendix F:
LINES THAT MEET GIVEN CONSTRAINTS

1. Through two points, (X_1, Y_1) and (X_2, Y_2)

There are many forms to represent the line equation. For this case, we will use the form

$$mX - Y + b = 0 \qquad \text{(F-1)}$$

where m is the slope of the line and b is the Y intercept. This is the slope-intercept form derived in analytic geometry texts. In terms of the given (input) data,

$$m = \frac{Y_2 - Y_1}{X_2 - X_1}$$

$$b = Y_1 - mX_1 \qquad \text{(F-2)}$$

To illustrate, we will find the line equation that passes through $(2, -1)$ and $(4, 5)$:

$$m = \frac{6}{2} = 3$$

$$b = -1 - (3)(2) = -7$$

Thus, we have

$$3X - Y - 7 = 0$$

2. Through a point with a given slope

We use formula (F-1) as follows. Suppose, for example, $m = -1.5$ and the point is $(-1, 4)$. Then

$$-1.5(-1) - 4 + b = 0$$

from Eq. (F-1), or

$$b = 2.5$$

We have

$$-1.5X - Y + 2.5 = 0$$

Multiplying by -2 gives

$$3X + 2Y - 5 = 0.$$

3. Line with a given slope and intercept

If we are given m and b, formula (F-1) is applied directly. If we are given m and the X intercept, a, we can use the method of paragraph 2 above; i.e.,

$$X_1 = a \quad \text{and} \quad Y_1 = 0.$$

4. Line through a point and parallel to a given line

If we use the line form,

$$AX + BY + C = 0 \qquad \text{(F-3)}$$

for the given line, then $m = -A/B$ is the slope for both the given and parallel lines. Then with this slope and the given point, we use the method of paragraph 2. To illustrate, suppose the given line is $4X + 3Y + 1 = 0$ and the point through which the parallel line passes is $(3, -2)$.

$$m = -\frac{4}{3}$$

$$-\frac{4}{3}(3) - (-2) + b = 0$$

$$b = 2$$

Thus

$$-\frac{4}{3}X - Y + 2 = 0$$

or

$$4X + 3Y - 6 = 0$$

is the equation of the parallel line.

5. Line through a point and perpendicular to a given line

The slope of a perpendicular line is the negative reciprocal of the slope of the given line. Thus using formula (F-3) and referring to paragraph 4, the slope of the perpendicular line will be B/A. This slope and the given point are used as in previous paragraphs to derive the equation.

6. Line through a point that makes an angle, θ, with either the X or Y axis

If θ is measured positive in the counterclockwise direction from the X axis, then $m = \tan \theta$. The given point and m form the input data to use the point/slope form of paragraph 2. If θ is measured counterclockwise from the Y axis, then

$$m = -\cot \theta \quad \text{or} \quad -\frac{1}{\tan \theta}$$

This slope and the given point are used as before.

7. Line at a given angle, θ, with respect to a given line and which passes through a given point

If the given line is in the form of Eq. (F-3), then $m = -A/B$. The angle of that line, α, is given by

$$\alpha = \tan^{-1} m$$

where α is positive or negative acute (less than 90°) according to the algebraic sign of m. Then, the angle of the desired line will be $\alpha + \theta$. The slope of the desired line will thus be

$$m = \tan(\alpha + \theta) \tag{F-4}$$

To illustrate, suppose the given line is $3X + 3Y - 2 = 0$; the desired line will make an angle of 75° with the given line and will pass through (6, 2). The slope of the given line is $-A/B = -1$. Hence

$$\alpha = \tan^{-1}(-1) = -45°$$

The desired angle then is $-45° + 75° = 30°$. From Eq. (F-4), $m = \tan 30° = \sqrt{3}/3 \approx 0.58$. Now from paragraph 2,

$$0.58(6) - 2 + b = 0 \quad \text{or} \quad b = -1.48$$

The desired equation then is

$$0.58X - Y - 1.48 = 0$$

8. Line parallel to a given line at a given distance

Using the *normal* form of the equation of a line as derived in analytic geometry, we can find the distance from a line to a point. In this case, any general point on the parallel line will do. We use Eq. (F-3), $AX + BY + C = 0$, as the given equation. Then from analytics

$$d = \frac{\pm(AX + BY + C)}{\pm\sqrt{A^2 + B^2}} \qquad \text{(F-5)}$$

The plus sign for d is used if the *desired* line is above the *given* line. Minus is used if the desired line is below the given line. The square root sign is either plus or minus according to the sign of the Y coefficient, B, of the given line. To illustrate, consider Fig. F1.

From formula (F-5),

$$2 = \frac{+(3X - 4Y - 3)}{-\sqrt{3^2 + 4^2}}$$

or upon simplification

$$3X - 4Y + 7 = 0$$

is the desired equation. Note that parallel lines have the same slopes and hence the same X, Y coefficients. They differ only in the constant, C.

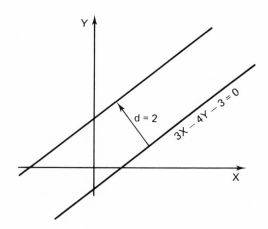

Figure F1 Equation of a line parallel to a given line at a given distance.

9. Line parallel to a given line at a given vertical distance from the given line

This case can be converted to that of paragraph 8 by converting the vertical distance to equivalent perpendicular distance. Figure F2 shows the relationship.

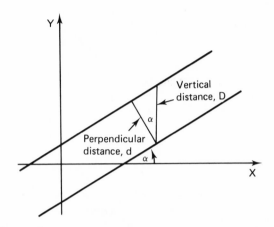

Figure F2 Deriving perpendicular distance from given distance, D.

The slope of the given line is $m = -A/B$ as described before. Then

$$\alpha = \tan^{-1} m$$

From Fig. F2 we see that the perpendicular distance, d, is

$$d = D \cos \alpha$$

Thus, this case now becomes reduced to that of paragraph 8.

10. Line from a given point, tangent to a given circle

The point is given as (X_1, Y_1) and the circle is given with radius, r, and center at (h, k). This is shown in the diagram of Fig. F3.

In this example, (X_1, Y_1) is (2, 2); (h, k) is (6, 1); r is 3. The first thing that is done to develop the simplest equations is to make a translation such that the center is at the origin. Thus,

$$X' = X - h$$
$$Y' = Y - k \qquad \text{(F-6)}$$

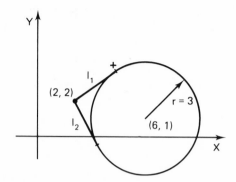

Figure F3 Example of line from a given point, tangent to a given circle.

Then

$$X_1' = X_1 - h$$
$$Y_1' = Y_1 - k$$

In our example this would make X_1' and Y_1' to be -4 and 0 respectively. After deriving the appropriate equation in $X'Y'$, we retranslate to put the equation in the original coordinate system. We wish to find m and b in order to use formula (F-1) for the line equation. We develop the solution for m and b by the following process. From Eq. (F-1), $-mX' + Y' - b = 0$. The distance from this line to the origin is r; the "normal" form of the line equation gives

$$\frac{-m(0) + (0) - b}{(m^2 + 1)^{1/2}} = r$$

or

$$r^2 = \frac{b^2}{m^{2+1}}$$

Since r is input, this is one equation which relates m and b. The other is

$$-mX_1' + Y_1' - b = 0$$

Solving simultaneously, we derive

$$(X_1'^2 - r^2)b^2 + 2Y_1'r^2b - (X_1'^2 + Y_1'^2)r^2 = 0$$

and

$$m = \frac{Y_1' - b}{X_1'} \qquad \text{(F-7)}$$

Formulas (F-7) apply only when the circle's center is at the origin. The first of the two equations is quadratic in terms of b and hence has two mathematical solutions, which are derived by the quadratic formula given in the text for formula (45). For each b there will be a corresponding m. However, the desired b (Y intercept) will be determined by the proximity to the Y intercept of the line joining (X_1, Y_1) to the *tracking symbol*—the $+$ of Fig. F3. Now we will see how the process works using the example of Fig. F3. Let us assume that the tracking symbol, placed visually near the point of tangency of the desired line, is located at (4, 5).

The translation given by (F-6) is:

$$X' = X - 6$$
$$Y' = Y - 1$$

Thus

$$X_1' = 2 - 6 = -4$$

and

$$Y_1' = 2 - 1 = +1$$

From Eq. (F-7), we get

$$\left[(-4)^2 - (3)^2\right]b^2 + 2(1)(3)^2 b - \left[(-4)^2 + (1)^2\right]3^2 = 0$$

$$7b^2 + 18b - 153 = 0$$

Solving the quadratic equation for b gives approximately $+3.6$ and -6.1. These are the solutions in the $X'Y'$ system. We note that the tracking symbol was placed at (4, 5) which is -2 and 4 in the $X'Y'$ system. Now the line from the point $(-4, 1)$ to $(-2, 4)$ in the $X'Y'$ system is derived by methods of paragraph 1. Its Y intercept is $+7$, which is closer to 3.6 than to -6.1. It is clear that the tracking symbol was not placed very well but nevertheless it did indicate the preferred b of 3.6. Now, from the other of the two Eqs. (F-7), we derive

$$m = \frac{1 - 3.6}{-4} = +0.65$$

The equation in $X'Y'$ is therefore

$$+ 0.65X' - Y' + 3.6 = 0$$

Retranslating we get

$$+0.65(X - 6) - (Y - 1) + 3.6 = 0$$

or

$$0.65X - Y + 0.7 = 0$$

which is the desired equation.

11. Line tangent to two given circles

There is a maximum of four lines that can be tangent to two circles. This is depicted in Fig. F4. In order to simplify the formulas to determine the lines, a translation and rotation are employed, as shown in Fig. F4. The derivation of the formulas to determine the lines is accomplished in the $X'Y'$ system and then transformations are applied to convert the lines to equations in the basic XY system.

In the $X'Y'$ system, one circle has coordinates $(0, 0)$ and the other has coordinates $(H, 0)$ where

$$H = \left[(h_2 - h_1)^2 + (k_2 - k_1)^2 \right]^{1/2} \tag{F-8}$$

We use the analytic geometry *normal* form of a line to get the distance from each line, r_1 and r_2, to the two centers. In getting these distances it is convenient to use the general line in the form,

$$X + BY + C = 0 \tag{F-9}$$

instead of the form of paragraph 1. (Either form requires the derivation of two parameters to find the equation.) With this form,

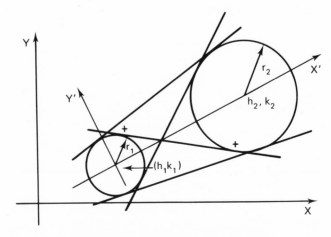

Figure F4   Line tangent to two circles.

the two equations that result for distances from line to point are:

$$\frac{C}{(1 + B^2)^{1/2}} = \pm r_1 \quad \text{and} \quad \frac{H + C}{(1 + B^2)^{1/2}} = \pm r_2 \quad \text{(F-10)}$$

H is known from application of Eq. (F-8) but B and C must be derived, so we ratio the two equations of (F-10) to get

$$\frac{C + H}{C} = \pm \frac{r_2}{r_1}$$

(The $\pm$ is necessary to cover all cases.) The two solutions for C are

$$C = \frac{r_1 H}{r_2 - r_1} \quad \text{and} \quad C = \frac{-r_1 H}{r_2 + r_1} \quad \text{(F-11)}$$

For each C, and using the first of the two Eqs. (F-10), we solve for B:

$$B = \pm \left(\frac{C^2 - r_1^2}{r_1^2} \right)^{1/2} \quad \text{(F-12)}$$

Formulas (F-11) and (F-12) give the four possible solutions in the $X'Y'$ system. At the graphic terminal, the operator has placed the tracking symbol "near" the two points of tangency of the preferred line. That line is selected automatically by the process of determining the sum of the absolute values of the two distances from each line to the tracking-symbol locations. The smallest sum represents the preferred line. The symbol locations are converted to the $X'Y'$ system prior to the distance test. The formulas of transformation for this, and also for converting the preferred line back to the XY system, are given as follows:

$$\theta = \tan^{-1} \left(\frac{k_2 - k_1}{h_2 - h_1} \right) \quad \text{(F-13)}$$

$$X' = X^* \cos \theta + Y^* \sin \theta \qquad Y' = -X^* \sin \theta + Y^* \cos \theta \quad \text{(F-14)}$$

and

$$X^* = X - h_1 \qquad Y^* = Y - k_1 \quad \text{(F-15)}$$

Algebraic simplification gives the desired equation.

As an illustration similar to the situation shown in Fig. (F-4), suppose we have two circles with centers at $(2, 2)$ and $(6, 5)$ with radii of one and three, respectively. We want to find the line that has the negative slope in the diagram; and so we visually place the tracking symbol, by inspection, near the points of tangency for that line (depicted on Fig. (F-4)). We will assume that they fall at (2.1, 2.9) and (5.8, 2.2). The line determination process is as follows: From Eq. (F-8),

$$H = \left[(6-2)^2 + (5-2)^2 \right]^{1/2} = 5$$

From Eq. (F-11)

$$C = \frac{1(5)}{3-1} = \frac{5}{2} \quad \text{and} \quad C = \frac{-1(5)}{3+1} = -\frac{5}{4}$$

From Eq. (F-12),

$$B = \pm \left[\frac{(+5/2)^2 - 1^2}{1^2} \right]^{1/2} \approx \pm 2.3$$

and

$$B = \pm \left[\frac{(-5/4)^2 - 1^2}{1^2} \right]^{1/2} = \pm 3/4$$

Now, using Eq. (F-9), we find that the four mathematical solutions in X'Y' are:

$$X' + 2.3\,Y' + \frac{5}{2} = 0$$

$$X' - 2.3\,Y' + \frac{5}{2} = 0$$

$$X' + \frac{3}{4}\,Y' - \frac{5}{4} = 0 \qquad\qquad \text{(F-16)}$$

$$X' - \frac{3}{4}\,Y' - \frac{5}{4} = 0$$

We convert the two tracking-symbol locations to the $X'Y'$ system as described before in order to get distances from lines to the points in a compatible system prior to making the final transformations. On translation, (2.1, 2.9) and (5.8, 2.2) become (0.1, 0.9) and (3.8, 0.2). The rotation angle, θ, from Eq. (F-13) is:

$$\theta = \tan^{-1} \frac{3}{4}$$

From this, we get $\cos \theta = 4/5$ and $\sin \theta = 3/5$. From Eq. (F-14) we now derive the $X'Y'$ locations; i.e.,

$$X' = 0.1\left(\frac{4}{5}\right) + 0.9\left(\frac{3}{5}\right) = +0.62$$

$$Y' = -0.1\left(\frac{3}{5}\right) + 0.9\left(\frac{4}{5}\right) = +0.66$$

and

$$X' = 3.8\left(\frac{4}{5}\right) + 0.2\left(\frac{3}{5}\right) = 3.16$$

$$Y' = -3.8\left(\frac{3}{5}\right) + 0.2\left(\frac{4}{5}\right) = -2.12$$

The distances from the first of the four Eqs. (F-16) to the two points are:

$$d_1 = \frac{0.62 + 2.3(0.66) + 5/2}{(1 + 2.3^2)^{1/2}} = 1.68$$

$$d_2 = \frac{3.16 + 2.3(-2.12) + 5/2}{(1 + 2.3^2)^{1/2}} = 0.32$$

$$|d_1| + |d_2| = 2.00$$

Similarly for the other three lines, the sums are

4.8, 0.36, and 3.7

Clearly, the third equation represents the line closest to the two tracking symbols and is the desired solution in the $X'Y'$ system. The correct solution in $X'Y'$ is also correct in XY, and so we now convert first to X^*Y^* by θ rotation using Eq. (F-14). Thus,

$$X' + \frac{3}{4}Y' - \frac{5}{4} - 0 \quad \text{or} \quad 4X' + 3Y' - 5 = 0$$

becomes

$$4\left[X^*\left(\frac{4}{5}\right) + Y^*\left(\frac{3}{5}\right)\right] + 3\left[-X^*\left(\frac{3}{5}\right) + Y^*\left(\frac{4}{5}\right)\right] - 5 = 0$$

or

$$\frac{7}{5}X^* + \frac{24}{5}Y^* - 5 = 0 \quad \text{or} \quad 7X^* + 24Y^* - 25 = 0$$

Finally, we translate back to XY by replacing X^* by $X - 2$ and Y^* by $Y - 2$. Thus, we get

$$7X + 24Y - 87 = 0$$

When the analytical solution can be achieved in this kind of straightforward way, it is recommended that the computer perform the operations in contrast to having all solutions displayed and the operator indicating the choice on the display.

12. Line from a given point, tangent to a given ellipse

Suppose we know the equation of an ellipse and wish to derive one of the two tangent lines from an external given point. If the ellipse has rotated axes and center at (h, k), it is suggested that an $X'Y'$ system be developed as in paragraph 11 such that the ellipse will be in standard position. The point is also converted to $X'Y'$ so that the desired line is determined in $X'Y'$ and then converted to XY as in paragraph 11.

The formulas that will be presented here will, of course, apply to a special ellipse which is a circle and, therefore, might be used in lieu of formula (F-7), developed in paragraph 10, for the circle. That formula is in a bit simpler form, however.

Now, the ellipse in standard form may be written as

$$\frac{X'^2}{C_1} + \frac{Y'^2}{C_2} = 1$$

and the given point in the $X'Y'$ system is (X_1', Y_1'). The following formulas apply (we will use X and Y instead of $X'Y'$ to simplify notation).

$$K_1 = C_1 Y_1^2 + C_2 X_1^2$$
$$K_2 = -2C_1 C_2 Y_1 \qquad \text{(F-17)}$$
$$K_3 = C_1 C_2^2 - C_2^2 X_1^2$$

$$K_1 Y^2 + K_2 Y + K_3 = 0 \qquad \text{(F-18)}$$

We solve for Y by quadratic formula as was done in the text for Eq. (45). Then,

$$X = \frac{C_1(C_2 - Y_1 Y)}{C_2 X_1} \qquad \text{(F-19)}$$

The two solutions are the points of tangency from the point to the ellipse. The preferred point is selected by placing the tracking symbol near to the solution of interest. The smaller of the two distances from the symbol to the two solutions is picked. That solution along with the given point are the two points from which the equation of the line is derived, as in paragraph 1.

As an example, suppose $C_1 = 9$, $C_2 = 4$, $X_1 = 5$, and $Y_1 = 4$. Then, from formulas (F-17) through (F-19),

$$K_1 = 244 \qquad K_2 = -288 \qquad K_3 = -256$$

$$Y = 1.77, \; -1.10$$

(We select $Y = 1.77$ as the preferred ordinate). For $Y = 1.77$, $X = 1.40$. These coordinates and (5, 4) will give the desired equation. Had there been any initial transformations, this would have to be considered in converting to the final equation.

The methods in paragraphs 1 through 12 give analytical methods for line development. For more complex functions where complete automation is impractical, the use of graphics to indicate visually the approximate point of tangency may suffice for many applications.

Appendix G:
SAMPLE PROGRAM
FOR BEZIER CURVE GENERATION[1]

```
      SUBROUTINE BEZIER(NPTS,XC,YC,ZC,NPOL,XP,YP,ZP)
      INTEGER FACT
C...
C...  THIS SUBROUTINE CALCULATES THE COORDINATES (XC,YC,ZC) FOR NPTS
C...  ON A BEZIER CURVE.  THE INPUT DATA ARE THE NPOL POINTS OF THE
C...  BEZIER POLYGON GIVEN IN (XP,YP,ZP).
C...
C...  THIS SUBROUTINE UTILIZES A FUNCTION FACT(N) WHICH COMPUTES THE
C...  THE FACTORIAL OF N.
C
      DIMENSION XC(NPTS),YC(NPTS),ZC(NPTS)
      DIMENSION XP(NPOL),YP(NPOL),ZP(NPOL)
C...
C...  THE PARAMETRIC ITERATION INTERVAL
      DELTAT=1.0/FLOAT(NPTS-1)
C...
      XC(1)=XP(1)
      YC(1)=YP(1)
      ZC(1)=ZP(1)
C...
C...  THE PARAMETER N (HERE NBEZ) IS ONE LESS THAN THE NUMBER OF
C...  POINTS ON THE POLYGON
      NBEZ=NPOL-1
C...
      NLIM=NPTS-1
C...
C...  WITH THE FORMULA ONE MAY NOT COMPUTE THE END POINTS DUE TO THE
C...  DEGENERATE FORM OF THE FACT FUNCTION.
C...
      DO 2000 INTEP=2,NLIM
         T=(INTEP-1)*DELTAT
         XC(INTEP)=0.
         YC(INTEP)=0.
         ZC(INTEP)=0.
C...
         DO 1000 J=1,NPOL
            I=J-1
            BEZBIN=FLOAT(FACT(NBEZ))/FLOAT(FACT(I)*FACT(NBEZ-I))
            BEZWT =BEZBIN*(T**(I))*((1.0-T)**(NBEZ-I))
C...
            XC(INTEP)=XC(INTEP)+XP(J)*BEZWT
            YC(INTEP)=YC(INTEP)+YP(J)*BEZWT
            ZC(INTEP)=ZC(INTEP)+ZP(J)*BEZWT
1000     CONTINUE
2000  CONTINUE
      XC(NPTS)=XP(NPOL)
      YC(NPTS)=YP(NPOL)
      ZC(NPTS)=ZP(NPOL)
      RETURN
      END
```

[1]This program is reprinted with the permission of Dr. Bertram Herzog, Director, University Computing Center, University of Colorado.

BIBLIOGRAPHY

Since there are many publications that relate in varying degrees to the theme of this text, considerable subjectivity was exercised in the selection of the following publications for further study. Much of the published literature is research-oriented in the sense that interesting techniques are developed regardless of known or presumed applicability. Also, some subjects, such as splines, are treated much more extensively than others independent of proportion of utility.

With these points in mind, the following criteria were used as guidelines in establishing this listing of additional reading for those who wish to expand their technical base for computer graphic applications. The publication should:

1. Present a technique, procedure, and/or computer program for the development of curves and/or surfaces.
2. Be applicable to interactive graphic applications whether or not it is explicitly treated.
3. Be usable for a broad range of applications.
4. Be of relatively basic mathematical foundation to enhance understanding by a large class of potential users.
5. Be of most recent vintage, for a particular author, when several documents represent successive updates of a specific subject.

ADAMS, J. A., "Geometric Concepts for Computer Graphics," Engineering and Weapons Report, No. EW-72-4, U.S. Naval Academy, Annapolis, Md., September 1972.

——, "A Comparison of Methods for Cubic Spline Curve Fitting," U.S. Naval Academy, Division of Engineering and Weapons, Annapolis, Md., August 1973.

———, "Cubic Spline Curve Fitting with Controlled End Conditions," *Computer Aided Design*, **6** (1974), 1–9.

AHLBERG, J. H., E. N. NILSON, and J. L. WALSH, *The Theory of Splines and Their Applications*. New York: Academic Press, 1967.

AHUJA, D. V., "An Algorithm for Generating Spline-like Curves," *IBM Systems Journal*, **7**, no. 3/4 (1968), 206–217.

AHUJA, D. V., and S. A. COONS, "Geometry for Construction and Display," *IBM Systems Journal*, **7**, no. 3/4 (1968), 188–205.

AKIMA, H., "A New Method of Interpolation and Smooth Curve Fitting Based on Local Procedures," *Journal of the ACM* (1970).

———, "A Method of Bivariate Interpolation and Smooth Surface Fitting Based on Local Procedures," *Communications of the ACM*, **17**, no. 1 (January 1974).

ALBASINY, E. L., and W. D. HOSKINS, "The Numerical Calculation of Odd-Degree Polynomial Splines with Equi-Spaced Knots," *Journal of the Institute of Mathematics Applications*, **7** (1971), 384–397.

AMBLER, A. P., and R. J. POPPLESTONE, "Turning Spatial Relations into Equations," Edinburgh University School of Artificial Intelligence Memo MIP-R-107, Edinburgh, Scotland, May 1974.

APPEL, A., "Modelling in Three Dimensions," *IBM Systems Journal* (1968).

ARMIT, A. P., "A Multipatch Design System for Coons' Patches," *IEEE Conference Publication*, no. 51 (April 1969), 152–161.

———, "Interactive 3D Shape Design—MULTIPATCH and MULTIOBJECT," *Curved Surfaces in Engineering*, ed. L. J. I. Browne, pp. 26–37. Guilford, England: IPC Science and Technology Press Ltd., 1972.

ARMIT, A. P., and A. R. FORREST, "Interactive Surface Design," *Advanced Computer Graphics: Economics, Techniques and Applications* (*Computer Graphics 70*), eds. R. D. Parslow, R. E. Green, pp. 1179–1202. London: Plenum Press, 1971.

ASKER, B., "The Spline Curve—A Smooth Interpolating Function Used in Numerical Design of Ship Lines," *Nord. Tidskr. Inform. Behandling*, **2** (1962), 76–82.

BAER, A., C. EASTMAN, and M. HENRION, "A Survey of Geometric Modelling," Research Report 66, Dept. of Architecture, Carnegie-Mellon University, Pittsburgh, Pa., March 1977.

BARNHILL, R. E., "Blending Function Interpolation: A Survey and Some New Results," Numerical Analysis Reports No. 9, Dept. of Mathematics, University of Dundee, Scotland, July 1975.

BARNHILL, R. E., "Representation and Approximation of Surfaces," *Mathematical Software III*, ed. J. R. Rice, pp. 68–119. New York: Academic Press, 1977.

BARNHILL, R. E., "Smooth Interpolation Over Triangles," *Computer Aided Design*, eds. R. E. Barnhill, R. F. Riesenfeld, pp. 45–70. New York: Academic Press, 1974.

BARNHILL, R. E., J. H. BROWN, and I. M. KLUCEWICZ, "A New Twist for Computer Aided Geometric Design," *Computer Graphics and Image Processing* (to appear).

BARNHILL, R. E., R. P. DUBE, G. J. HERRON, F. F. LITTLE, and R. F. RIESENFELD, "Representation and Approximation of Surfaces" (a movie), 1977.

BARNHILL, R. E., and J. A. GREGORY, "Blending Function Interpolation to Boundary Data on Triangles," Technical Report TR/14, Dept. of Math., Brunel Univ., Uxbridge, England, 1972.

BARNHILL, R. E., and J. A. GREGORY, "Smooth Polynomial Interpolation to Boundary Data on Triangles," Technical Report TR/31, Dept. of Math., Burnel Univ., Uxbridge, England, 1973.

BARNHILL, R. E. and J. A. GREGORY, "Polynomial Interpolation to Boundary Data on Triangles," *Math. Comp.*, **29** (1975), 726–235.

BARNHILL, R. E., and R. F. RIESENFELD, eds., *Computer Aided Design* (proceedings of a conference held at The University of Utah, Salt Lake City, Utah, 18–21 March 1974). New York: Academic Press, 1974.

BAUMGART, B. G., "Geometric Modelling for Computer Vision," Memo AIM-249, STAN-CS-74-463, Stanford Artificial Intelligence Laboratory, Stanford, Ca., October 1974.

BEZIER, P., *Numerical Control—Mathematics and Applications*, trans. A. R. Forrest. London: John Wiley & Sons, 1972.

_____, "Mathematical and Practical Possibilities of UNISURF," *Computer Aided Geometric Design*, eds. R. E. Barnhill, R. F. Riesenfeld, pp. 127–157. New York: Academic Press, 1974.

_____, "Raccordement Automatique des Courbes et des Surfaces Parametriques," Renault PB/MG Sce 0800, June 1974. Boulgne-Billancourt, France.

_____, "Degenerate Surfaces and Particularly About Three-Sided Patches," Renault, April 1975. Boulgne-Billancourt, France.

BIRKHOFF, G., "Nonlinear Interpolation by Splines, Pseudosplines, and Elastica," General Motors Research Laboratories (Warren, Michigan), GMR468, February 1965.

_____, "The Draftsman's and Related Equations," General Motors Research Laboratories (Warren, Michigan), GMR748, March 1968.

_____, "Piecewise Bicubic Interpolation and Approximation in Polygons," *Approximation with Special Emphasis on Spline Functions*, ed. Schoenberg. New York: Academic Press, 1969.

BOYSE, J. W., "Interference Detection Among Solids and Surfaces," General Motors Research Laboratories (Warren, Michigan), GMR-2470, July 1977.

BRAID, I. C., *Designing with Volumes*. Cambridge, England: Cantab Press, 1974.

_____, "Six Systems for Shape Design and Representation—a Review," Cambridge University (Cambridge, England) CAD Group Doc. 87, May 1975.

_____, "A New Shape Design System," Cambridge University (Cambridge, England) CAD Group Doc. 89, March 1976.

BROWN, I. J., ed., *Curved Surfaces in Engineering*. Guilford, England: IPC Science and Technology Press, Ltd., 1972.

BUTTERFIELD, K. R., "The Computation of All the Derivatives of a B-Spline Basis," British Leyland, Oxford, England, May 1974. (To appear, *Journal of the Institute of Mathematics Applications*.)

CALU, J. V. J., "Implementation of Surface Interpolation for Computer Graphics," M.S. Thesis, University of Utah, Salt Lake City, Utah, December 1974.

CATMULL, E., and R. ROM, "A Class of Local Interpolating Splines," *Computer Aided Geometric Design*, eds. R. E. Barnhill, R. F. Riesenfeld, pp. 317–326. New York: Academic Press, 1974.

CHAIKIN, G. M., "Geometric Description and Generation of Surfaces," Report No. CRL-348, AD AO32692, Div. of Applied Science, New York Univ., New York, January 1975.

CHASEN, S. H., "Computer Aided Design Concepts," First AIAA Summer Institute on Aircraft Design (Theme: Special Applications of Computers and Graphics to Aircraft Design), Marietta, Ga., 15–17 August 1977.

CLARK, J. H., "Designing Surfaces in 3-D" *Comm. ACM*, Vol. 19, No. 8, August 1976.

CLINE, A. K., "Scalar- and Planar-Valued Curve Fitting Using Splines under Tension," *Comm. ACM*, **17**, no. 4 (1974), 218–220.

COONS, S. A., *Surfaces for Computer Aided Design*, Design Division, Mech. Engin. Dept., Massachusetts Institute of Technology, Cambridge, Mass., 1967 (available as AD663 504 from NTIS, Springfield, Virginia).

_____, *Surfaces for Computer Aided Design of Space Forms*, MAC-TR-41, Project MAC, Massachusetts Institute of Technology, Cambridge, Mass., June 1967.

_____, "Surface Patches and Piecewise Blending Functions," Systems and Information Science, Syracuse University, Syracuse, N.Y., June 1972.

_____, "Surface Patches and B-Spline Curves," *Computer Aided Geometric Design*, eds. R. E. Barnhill, R. F. Riesenfeld. New York: Academic Press, 1974.

COX, M. G., "Cubic-Spline Fitting with Convexity and Concavity Constraints," DNAC 23, National Physical Laboratory, Teddington, England.

_____, "The Numerical Evaluation of a Spline from its B-Spline Representation," NAC 68, National Physical Laboratory, Teddington, England, July 1976.

_____, "The Incorporation of Boundary Conditions in Spline Approximation Problems," NAC 80, National Physical Laboratory, Teddington, England, June 1977.

DEBOOR, C., "Bicubic Spline Interpolation," *J. Math. Phys.*, **41** (1962), 212–218.

_____, "On Calculating with B-Splines," *J. Approx. Theory*, **6** (1972), 50–62.

DIMSDALE, B., and K. JOHNSON, "Multiconic Surfaces," *IBM Journal of Research and Development* (November 1975), 523–529.

DUBE, R. P., "Local Schemes for Computer Aided Geometric Design," Ph.D. Thesis, University of Utah, Salt Lake City, Utah, 1975.

_____, "Univariate Blending Functions and Alternatives," *Computer Graphics and Image Processing*, **6** (1977), 394–408.

DUBE, R. P., G. J. HERRON, F. F. LITTLE, and R. F. RIESENFELD, "SURFED—An Interactive Editor for Free-Form Surfaces," Submitted to *Computer Aided Design*, 1977.

DUCHON, J., "Functions-Spline a Energie Invariante par Rotation," Research Report No. 27, Applied Mathematics and Informatics, University of Grenoble, Grenoble, France, January 1976.

EASTMAN, C. M., "The Concise Structuring of Geometric Data for Computer Aided Design," *Data Structures, Computer Graphics, and Pattern Recognition*, eds. A. Klinger, et al. New York: Academic Press, 1977.

ELLIS, T. M. R., and D. H. MCLAIN, "Algorithm 514: A New Method of Cubic Curve Fitting Using Local Data (E2)," *ACM Trans. on Mathematical Software*, **3**, no. 2 (June 1977).

FERGUSON, J. C., "Multivariable Curve Interpolation," *J. ACM*, **11**, no. 2 (April 1964), 221–228.

FLANAGAN, D. L., and O V. HEFNER, "Surface Molding—New Tool for Engineer," *Astronautics and Aeronautics*, **5**, no. 4 (April 1967), 58–62.

FORREST, A. R., "Curves and Surfaces for Computer Aided Design," Cambridge University (Cambridge, England) CAD Group, Ph.D Thesis, July 1968, 132 pp.

_____, "Analytic Curves for Describing Engineering Shapes," Specialist Session, International Computer Graphics Symposium, Brunel Univ., Uxbridge, England, July 1968.

_____, "Curves for Computer Graphics," *Pertinent Concepts in Computer Graphics*, eds. M. Faiman, J. Nievergelt, pp. 31–47. Urbana, Ill.: University of Illinois Press, 1969.

———, "On Coons' Surfaces and Bivariate Functional Interpolation," Cambridge University CAD Group Doc. 26, May 1969.

———, "The Twisted Cubic Curve," Cambridge University CAD Group Doc. 50, November 1970.

———, "Interactive Interpolation and Approximation by Bezier Polynomials," *Computer Journal*, **15**, no. 1 (January 1972), 71–79.

———, "On Coons' and Other Methods for the Representation of Curved Surfaces," *Computer Graphics and Image Processing*, **1** (1972), 341–359.

———, "A New Curve Form for Computer-Aided Design," Cambridge University CAD Group Doc. 66, June 1972.

———, "Computational Geometry—Achievements and Problems," *Computer Aided Geometric Design*, eds. R. E. Barnhill, R. F. Riesenfeld. New York: Academic Press, 1974.

GEARY, S., "A Computer Graphics System to Aid the Study of Interpolants to Curved Surfaces," M.S. Thesis, University of Utah, Salt Lake City, Utah, August 1973.

GORDON, W. J., "Spline-blended Surface Interpolation through Curve Networks," *J. Math. Mech.*, **18**, no. 10 (1969), 931–952.

———, "Free-Form Surface Interpolation through Curve Networks," General Motors Research Publication, GMR-921, October 1969, (Synopsis in Proceedings of the 1969 SIAM-ACM Conference on Mathematical Aids to Computerized Design).

———, "Blending-Function Methods of Bivariate and Multivariate Interpolation and Approximation," *SIAM Journal on Numerical Analysis*, **8**, no. 1 (March 1971), 158–177.

GORDON, W. J., and G. BIRKHOFF, "The Draftsman's and Related Equations," *J. On Approximation Theory*, **1**, (1968), 199–208.

GORDON, W. J., R. E. BARNHILL, and G. BIRKHOFF, "Smooth Interpolation in Triangles," *Journal of Approximation Theory*, **8**, no. 2 (June 1973), 114–128.

GORDON, W. J., and R. F. RIESENFELD, "Bernstein-Bezier Methods for the Computer Aided Design of Free-Form Curves and Surfaces," *Journal of the ACM*, **21**, no. 2, (April 1974), 293–310.

GORDON, W. J., and R. F. RIESENFELD, "B-Spline Curves and Surfaces," *Computer Aided Geometric Design*, eds. R. E. Barnhill, R. F. Riesenfeld, pp. 95–126. New York: Academic Press, 1974.

GRANVILLE, P. S., "Geometric Characteristics of Streamlined Shapes," *Journal of Ship Research*, **13**, no. 4 (December 1969).

GREGORY, J. A., "Smooth Interpolation without Twist Constraints," *Computer Aided Geometric Design*, eds. R. E. Barnhill, R. F. Riesenfeld, pp. 71–87. New York: Academic Press, 1974.

GREVILLE, T. N. E., ed., *Theory and Applications of Spline Functions*. New York: Academic Press, 1969.

_____, "Numerical Procedures for Interpolation by Spline Functions," *J. SIAM, Numerical Anal. Ser. B*, **1** (1964), 53–68.

HALL, C. A., "Natural Cubic and Bicubic Spline Interpolation," *SIAM Numerical Analysis*, **10** (December 1973).

HALL, S. J., "Curve Fitting of Badly Spaced Data Using Uniform Basis B-Splines," Hawker-Siddeley Aviation, Brough, (W. Morland, England) YBP 3272, October 1973.

HARTLEY, P. J., and C. J. JUDD, "Parametrisation of Bezier Type B-Spline Curves and Surfaces," Dept. of Mathematics, Lanchester Polytechnic, 1977.

HAYES, J. G., "New Shapes from Bicubic Splines," NSAC 58, National Physical Laboratory, Teddington, England, September 1974.

_____, "Numerical Methods for Curve and Surface Fitting," NAC 50, National Physical Laboratory, Teddington, England, April 1974.

HERZOG, B., and G. VALLE, "Interactive Control of Surface Patches (from a Remote Graphic Terminal)," Rapporto N. 1, Instituto di Elettronica, Universita di Bologna, Bologna, Italy, April 1970.

ICHIDA, K., T. KIYON, and F. YOSHIMOTO, "Curve Fitting by a One-Pass Method with a Piecewise Cubic Polynomial," *ACM Trans. on Mathematical Software*, **3**, no. 2 (June 1977).

JOHNSON, R. S., A. L. FULLER, J. A. CLAFFEY, F. R. BJORKLAND, and B. M. THOMSON, "Man-Computer Graphics in Preliminary Ship Design," AIAA/SNAME meeting, Annapolis, Md., July 1972.

JOHNSON, T. E., "Arbitrarily Shaped Space Curves for C.A.D.," Summer session course on Computer-Aided Design, Massachusetts Institute of Technology, Cambridge, Mass., 1–12 August 1966.

JOHNSON, W. L., "Analytic Surfaces for Computer-Aided Design," Paper 660152, Society of Automative Engineers Convention Proceedings (Detroit, Michigan) January 10–14, 1966.

KELLY, J. P., and N. E. SOUTH, "Analytic Surface Methods," Numerical Control Development Unit Ford Motor Co., Dearborn, Mich., December 1965.

KUIPER, G., "Preliminary Design of Ships Lines by Mathematical Methods," *Journal of Ship Research*, **14**, no. 1 (March 1970), 52–66.

LANE, J. M., and R. F. RIESENFELD, "The Application of Total Positivity to Computer-Aided Curve and Surface Design," Dept. of Computer Science, University of Utah, Salt Lake City, Utah, 1977.

LAVICK, J. J., and G. L. MARTIN, "Modern Techniques in Design," 1972 CAD/CAM Conference, Society of Manufacturing Engineers, Atlanta, Ga., February 1972.

LAWSON, D. L., "C1-Compatible Interpolation over a Triangle," Computing Memo 407, Jet Propulsion Lab., California Institute of Technology, Pasadena, Ca., March 1976.

"Lockheed Computer Graphics Augmented Design and Manufacturing System," CADAM, Lockheed-California Co., Burbank, Ca.

LYCHE, T. and L. L. SCHUMAKER, "Algorithm 480: Procedure for Computing, Smoothing, and Interpolating Natural Splines (E1)," *Comm. ACM*, **17**, no. 8 (August 1974).

———, "Computation of Smoothing and Interpolating Natural Splines via Local Bases," *SIAM J. Numer. Anal.*, **10**, no. 6 (December 1973).

MACCALLUM, K. J., "Surfaces for Interactive Graphical Design," *Computer Journal*, **13**, no. 4 (November 1970), 352–358.

———, "Mathematical Design of Hull Surfaces," Royal Institution of Naval Architecture, December 1971.

———, "The Use of Interactive Graphics for the Preliminary Design of a Ship's Hull," *Advanced Computer Graphics: Economics, Techniques and Applications (Computer Graphics 70)*, eds. R. D. Parslow, R. E. Green, pp. 1203–1216. London: Plenum Press, 1971.

MANNING, J. R., "C-Spline Interpolation," Shoe and Allied Trades Research Association, 1974.

MCALLISTER, D. F., E. PASSOW, and J. A. ROULIER, "Algorithms for Computing Shape Preserving Spline Interpolations to Data," *Maths. of Computation*, **31**, no. 139 (July 1977), 717–725.

MCLAIN, D. H., "Drawing Contours from Arbitrary Data Points," *Computer Journal*, **17**, no. 4 (November 1974), 318–324.

MEHLUM, E., "A Short Description of the Mathematical Fairing of a Single Curve, Being the Basis of the Autokon Hull Fairing Program," Oslo, Sweden, September 1969.

———, " 'BOFOS' Program for Automobile and Aircraft Surfaces," Centralinstitutt for Industriell Forskning, Oslo, Sweden, December 1969.

———, "Curve and Surface Fitting Based on Variational Criteria for Smoothness," Centralinstitutt for Industriell Forskning, Oslo, Sweden, December 1969.

———, "Nonlinear Splines," *Computer Aided Geometric Design,* eds. R. E. Barnhill, R. E. Riesenfeld, pp. 173–207. New York: Academic Press, 1974.

"New Three-Dimensional Modelling Technique," *Bell Labs Record*, **49**, no. 8 (September 1971), Bell Telephone Laboratories.

NEWELL, A., "A General Discussion of the Use of Conic Equations to Define Curved Surfaces," Doc. D2-4398, The Boeing Company, March 1960.

NEWELL, M., "VINO—A New Approach to Coordinate Transformations and Graphical Modelling," CAD Centre, Cambridge, England.

NIELSON, G. M., "Some Piecewise Polynomial Alternatives to Splines Under Tension," *Computer Aided Geometric Design*, eds. R. E. Barnhill, R. F. Riesenfeld, pp. 209–235. New York: Academic Press, 1974.

_____, "Computation of v-Splines," Technical Report NR 044-443-11, Dept. of Mathematics, Arizona State University, Tempe, Ariz., 1974.

NUTBOURNE, A. W., R. B. MORRIS, and C. M. HOLLINS, "A Cubic Spline Package: Part 1—User's Guide," *Computer Aided Design*, **4**, no. 5 (1972).

_____, "A Cubic Spline Package: Part 2—The Mathematics," *Computer Aided Design*, **5**, no. 1 (1973), 3–13.

PALMER, T. R., "Shape Representation in Computer-Aided Design," Computational Geometry Project Memo. CGP77/5, August 1977.

"Parametric Surfaces in a Computer Graphics Design System," Lockheed California Co. Report LR 26059, 41-5610-3891, Burbank, Ca., January 1974.

PAYNE, P. J., "Contouring Program for Coons' Surface Patches," Cambridge University (Cambridge, England), CAD Group Doc. 16, November 1968.

_____, "A Contouring Program for Joined Surface Patches," Cambridge University (Cambridge, England), CAD Group Doc. 58, June 1971.

PETERS, G. J., "Interactive Computer Graphics Application of the Parametric Bi-Cubic Surface to Engineering Design Problems," *Computer Aided Geometric Design*, eds. R. E. Barnhill, R. F. Riesenfeld, pp. 259–302. New York: Academic Press, 1974.

PILCHER, D. T., "Smooth Parametric Surfaces," *Computer Aided Geometric Design*, pp. 237–253. New York: Academic Press, 1974.

_____, "Smooth Approximation of Parametric Curves and Surfaces," Ph.D. Thesis, University of Utah, Salt Lake City, Utah, 1973.

POEPPELMEIER, C. C., "A Boolean Sum Interpolation Scheme to Random Data for Computer Aided Geometric Design," M.Sc. Thesis, Dept. of Computer Science, University of Utah, Salt Lake City, Utah, December 1975.

POSDAMER, J. L., "A Vector Development of the Fundamentals of Computational Geometry," *Computer Graphics and Image Processing*, **6** (1977), 382–393.

_____, "Working Draft of a Paper on the Complexity of the Hidden Surface Problem," Dept. of Computing and Information Science, Syracuse Univ., Syracuse, N.Y., 1977.

RIESENFELD, R. F., "Application of B-Spline Approximation to Geometric Problems of Computer Design," University of Utah UTEC-CSs-73-126, Salt Lake City, Utah, March 1973.

_____, "Aspects of Modelling in Computer Aided Geometric Design," *Proc. of NCC*, **44**, 597–602, AFIPS Press (1975).

ROGERS, D. F., and J. A. ADAMS, *Mathematical Elements for Computer Graphics.* New York: McGraw-Hill, 1976.

SABIN, M. A., "The Representation of Shape for Computer-Aided Design," BAC, Weybridge, England.

_____, "Parametric Surfaces Equations for Non-Rectangular Regions," BAC, Weybridge (England), VTO/MS/147, July 1968.

_____, "Offset Parametric Surfaces," BAC, Weybridge (England), VTO/MS/149, September 1968.

_____, "Parametric Splines in Tension," BAC, Weybridge (England), VTO/MS/160, July 1970.

_____, "The Use of Circular Arcs to Form Curves Interpolated Through Empirical Data Points," BAC, Weybridge (England), VTO/MS/164, February 1971.

_____, "Trinomial Basis Functions for Interpolation in Triangular Regions (Bezier Triangles)," BAC, Weybridge (England), VTO/MS/188. July 1971.

_____, "Examination of a New Class of Spline Curves," BAC, Weybridge (England), VTO/MS/205, September 1974.

_____, "A Bezier-Like Surface Definition Controlled by Points Joined in an Arbitrary Network," Kongsberg Ltd., September 1976.

_____, "The Use of Piecewise Forms for the Numerical Representation of Shape," Report 60/1977, Computer and Automation Institute, Hungarian Academy of Science, Budapest, Hungary, 1977.

SCHWEIKERT, D. G., "An Interpolation Curve Using a Spline Under Tension," *J. Math. and Physics*, **45** (1966), 312–317.

SHAMOS, M. I., "Introduction to Computational Geometry," Dept. of Computer Science, Yale Univ., New Haven, Conn., October 1974.

"Shapes Manual—Introduction," Draper Lab., Massachusetts Institute of Technology, Cambridge, Mass., 1974.

SHEPARD, D., "A Two-Dimensional Interpolation Function for Irregularly Spaced Data," *Proc. 23rd Nat. Conf. ACM* (1965), 517–523.

SHIPPEY, G. A., "Parametric Representation of a Conic in Homogeneous Coordinates," Ferranti, Crewe Toll, Edinburgh TN/73/14, February 1973.

SHU, H., "Synthesis of 3-D Objects Having Complex Surface Boundaries," IITRI, Management and Computer Sciences Division (Chicago, Ill.), 1976.

STABLER, E. P., "Reconstruction of 3-Dimensional Images from 2-Dimensional Projections by a Transform Technique," Dept. of Electrical Engineering, Syracuse University, Syracuse, N. Y.

SUTHERLAND, I. E., R. F. SPROULL, and R. A. SCHUMAKER, "A Characterization of Ten Hidden-Surface Algorithms," *ACM Computing Surveys*, **6**, no. 1 (1974), 1–56.

WEILINGA, R. F., "Constrained Interpolation Using Bezier Curves as a New Tool in Computer Aided Geometric Design," *Computer Aided Geometric Design*, eds. R. E. Barnhill, R. F. Riesenfeld, pp. 153–172. New York: Academic Press, 1974.

WIXON, J. A., "Blending-Function Interpolation Over Non-Rectangular Domains," General Motors Corp. (Warren, Michigan), GMR-1957, September 1975.

WU, S-C, J. F. ABEL, and D. P. GREENBERG, "An Interactive Computer Graphics Approach to Surface Representation," *Comm. ACM*, **20**, no. 10 (October 1977).

YAMAGUCHI, F., "A Design System for Free Form Objects (FREEDOM), Technical Research Institute, Japan Society for the Promotion of Machine Industry, 1976.

ZWART, P. B., "Multivariate Splines with Non-Degenerate Partitions," *SIAM J. Numer. Anal.*, **10**, no. 4 (September 1973).

INDEX